Lecture Notes Editorial Policies

Lecture Notes in Statistics provides a format for the informal and quick publication of monographs, case studies, and workshops of theoretical or applied importance. Thus, in some instances, proofs may be merely outlined and results presented which will later be published in a different form.

Publication of the Lecture Notes is intended as a service to the international statistical community, in that a commercial publisher, Springer-Verlag, can provide efficient distribution of documents that would otherwise have a restricted readership. Once published and copyrighted, they can be documented and discussed in the scientific literature.

Lecture Notes are reprinted photographically from the copy delivered in camera-ready form by the author or editor. Springer-Verlag provides technical instructions for the preparation of manuscripts. Volumes should be no less than 100 pages and preferably no more than 400 pages. A subject index is expected for authored but not edited volumes. Proposals for volumes should be sent to one of the series editors or addressed to "Statistics Editor" at Springer-Verlag in New York.

Authors of monographs receive 50 free copies of their book. Editors receive 50 free copies and are responsible for distributing them to contributors. Authors, editors, and contributors may purchase additional copies at the publisher's discount. No reprints of individual contributions will be supplied and no royalties are paid on Lecture Notes volumes. Springer-Verlag secures the copyright for each volume.

Series Editors:

Professor P. Bickel
Department of Statistics
University of California
Berkeley, California 94720
USA

Professor P. Diggle
Department of Mathematics
Lancaster University
Lancaster LA1 4YL
England

Professor S. Fienberg
Department of Statistics
Carnegie Mellon University
Pittsburgh, Pennsylvania 15213
USA

Professor K. Krickeberg
3 Rue de L'Estrapade
75005 Paris
France

Professor I. Olkin
Department of Statistics
Stanford University
Stanford, California 94305
USA

Professor N. Wermuth
Department of Psychology
Johannes Gutenberg University
Postfach 3980
D-6500 Mainz
Germany

Professor S. Zeger
Department of Biostatistics
The Johns Hopkins University
615 N. Wolfe Street
Baltimore, Maryland 21205-2103
USA

Lecture Notes in Statistics

172

Edited by P. Bickel, P. Diggle, S. Fienberg, K. Krickeberg,
I. Olkin, N. Wermuth, and S. Zeger

Springer
New York
Berlin
Heidelberg
Hong Kong
London
Milan
Paris
Tokyo

Sneh Gulati
William J. Padgett

Parametric and Nonparametric Inference from Record-Breaking Data

Springer

Sneh Gulati
Department of Statistics
Florida International University
Miami, FL 33199
USA
gulati@fiu.edu

William J. Padgett
Department of Statistics
University of South Carolina
Columbia, SC 29208
USA
padgett@stat.sc.edu

Library of Congress Cataloging-in-Publication Data
Gulati, Sneh.
   Parametric and nonparametric inference from record-breaking data / Sneh Gulati,
William J. Padgett.
       p.   cm. — (Lecture notes in statistics ; 172)
   Includes bibliographical references and index.
   ISBN 0-387-00138-7 (softcover : alk. paper)
     1. Mathematical statistics.   2. Nonparametric statistics.   3. World records.   I. Padgett,
William J.   II. Title.   III. Lecture notes in statistics (Springer-Verlag) ; v. 172.
QA276 .G795   2003
519.5—dc21   2002042669

ISBN 0-387-00138-7   Printed on acid-free paper.

9 8 7 6 5 4 3 2 1   SPIN 10900284

Typesetting: Pages created by the authors using a Springer TEX macro package.

www.springer-ny.com

Springer-Verlag   New York Berlin Heidelberg
A member of BertelsmannSpringer Science+Business Media GmbH

Preface

As statisticians, we are constantly trying to make inferences about the underlying population from which data are observed. This includes estimation and prediction about the underlying population parameters from both complete and incomplete data. Recently, methodology for estimation and prediction from incomplete data has been found useful for what is known as "record-breaking data," that is, data generated from setting new records. There has long been a keen interest in observing all kinds of records—in particular, sports records, financial records, flood records, and daily temperature records, to mention a few. The well-known *Guinness Book of World Records* is full of this kind of record information. As usual, beyond the general interest in knowing the last or current record value, the statistical problem of prediction of the next record based on past records has also been an important area of record research. Probabilistic and statistical models to describe behavior and make predictions from record-breaking data have been developed only within the last fifty or so years, with a relatively large amount of literature appearing on the subject in the last couple of decades. This book, written from a statistician's perspective, is not a compilation of "records," rather, it deals with the statistical issues of inference from a type of incomplete data, *record-breaking data*, observed as successive record values (maxima or minima) arising from a phenomenon or situation under study.

Prediction is just one aspect of statistical inference based on observed record values. Estimation of unknown parameters of statistical or probabilistic models fitted to observed records, or testing hypotheses about one or more such parameters, is as important as prediction. In addition, nonparametric models for record data and the associated inference procedures have been developed over the past decade and a half. This monograph concerns all of these types of inference based on record-breaking data. A few other books have been written on various aspects of records, and with the exception of the recent book by Arnold, Balakrishnan, and Nagaraja (1998), they have focused mostly on stochastic behavior of records, characterization, and prediction. Arnold et al. gave an excellent comprehensive review of most of the results on records, including some material on inference from record-breaking data. However, their chapter on inference from such data is somewhat brief, discussing mainly the estimation of parameters and not the more general problem of parametric and nonparametric inference from record-breaking data.

The main purpose of the present monograph is therefore to fill the inference gap mentioned above. Although inferences for parametric models are presented briefly, along with a summary of some of the results on prediction of future records, we focus on cataloging the results on nonparametric inference from record-breaking data. Included is some material on parametric and nonparametric Bayesian models as well as a discussion of trend models. As a result, the material presented in this monograph provides a good supplement on inference from record-breaking data for a graduate course on the general topic of records. It should also serve as a valuable reference work on this topic for statisticians, probabilists, mathematicians, and other researchers.

The inference material presented here is intended to be somewhat self-contained, with many of the proofs and derivations of the important results included. For the basic estimation, prediction, and hypothesis testing results, a knowledge of at least first-year graduate-level mathematical statistics is assumed. For a good understanding of the asymptotic results on nonparametric function estimation in Chapters 4 and 5, basic knowledge of graduate-level probability theory and standard stochastic convergence theory is assumed.

After a brief introduction and background on stochastic characterizations of record values, Chapter 3 covers in some detail parametric inference from record-breaking data observed from exponential distributions and Weibull distributions. In addition, a summary of some of the results in the literature on prediction of future record values is given there for completeness. Chapter 4 begins the ideas of nonparametric inference procedures based on maximum likelihood methods and smooth nonparametric function estimation is described in detail in Chapter 5. Bayesian methods of inference are covered in Chapter 6, followed in Chapter 7 by models taking trends into account. The material in each chapter, after the background in Chapter 2, is somewhat independent, except that Chapters 4 and 5 should be covered together since the limiting behavior of the smooth function estimators relies on the asymptotic results in Chapter 4.

The authors express their gratitude to the Departments of Statistics at Florida International University and the University of South Carolina for their support. Particular thanks go to Professor Mounir Mesbah and the Department of Statistics at the University of South Brittany, France, for their support of the first author during her sabbatical leave when much of the drafting of these notes was accomplished. The second author was also partially supported by the National Science Foundation during the writing. In addition, thanks go to John Kimmel at Springer, the series editors, and the reviewers of the draft manuscript for their comments and suggestions on improving the presentation, clarity, and format of the material. Finally, our heartfelt appreciation is extended to our spouses and families for their support and patience during the preparation of this manuscript.

Sneh Gulati
William J. Padgett

Contents

Preface   v

1. Introduction   1

2. Preliminaries and Early Work   5
   2.1 Notation and Terminology   6
   2.2 Stochastic Behavior   7

3. Parametric Inference   11
   3.1 General Overview   11
   3.2 Parametric Inference Due to Samaniego and Whitaker   12
       3.2.1 Exponential Distribution—Inverse Sampling   12
       3.2.2 Exponential Distribution—Random Sampling   14
   3.3 Parametric Inference Beyond Samaniego and Whitaker   17
   3.4 The Problem of Prediction   24
       3.4.1 Best Linear Unbiased Prediction   25
       3.4.2 Best Linear Invariant Prediction   28
       3.4.3 "Other" Optimal Predictors   29
       3.4.4 Prediction Intervals   30

4. Nonparametric Inference—Genesis   33
   4.1 Introduction   33
   4.2 The Work of Foster and Stuart   33
   4.3 Nonparametric Maximum Likelihood Estimation   36
   4.4 Asymptotic Results   39

5. Smooth Function Estimation   45
   5.1 The Smooth Function Estimators—Definitions and Notation   46
   5.2 Asymptotic Properties of the Smooth Estimators   49

6. Bayesian Models   67
   6.1 Bayesian Prediction Intervals   67
       6.1.1 One-Parameter Exponential Model   68
       6.1.2 Two-Parameter Exponential Model   68
   6.2 Laplace Approximations for Prediction   69
   6.3 Bayesian Inference for the Survival Curve   73

7. Record Models with Trend   81
   7.1 Introduction   81
   7.2 The Models for Records with Trend   82

7.2.1 The F^α Model   82
7.2.2 The Pfeifer Model   83
7.2.3 The Linear Drift Model   84
7.3 The Geometric and the Pfeifer Models   84
7.3.1 The Geometric Model   84
7.3.2 The Pfeifer Model   87
7.4 Properties of the Linear Drift and Related Models   89
7.4.1 Early Work   89
7.4.2 Trend Models—The Work of Smith (1988) and Other
      Developments   95
7.5 The "General Record Model"   100

References   105

Index   111

1
Introduction

Everyone is interested in records, weather records, sports records, crime statistics, and so on. Record values are kept for almost every conceivable phenomenon. What was the coldest day last year (or ever), which city has the lowest crime rate, what was the shortest time recorded to complete a marathon, who holds the record in eating the most number of hot dogs in the shortest period, what was the highest stock value thus far? The list could go on and on; there is even a book that lists all kinds of records broken during a given year—the well-known *Guinness Book of World Records*! Naturally, if there is a subject concerning statistical values that interests the majority of people in the world, it has to be of interest to statisticians. However, how does one relate record values to statistical theory? The easiest way to explain this is with some examples. To begin with, consider a sports event: Not only do we want to know who holds the record for running 100 meters in the Olympics, but we also want to *predict* the next record-breaking time. Similarly, we want to determine if Miami, Florida, will still have the highest auto theft rate next year, or will Los Angeles still be the most polluted city in the US next year. Or we would like to predict the next highest closing price of a particular stock. In all of these examples, we want to use past data to predict the future. And prediction of the future using past data requires statistical theory.

Besides arising naturally in our day-to-day activities, observing record values also has a place in destructive stress testing and industrial quality control experiments. In these experiments it is often of interest to estimate a guarantee value or a population quantile. Generally, we would do this by observing the entire sample and then using the appropriate order statistic to estimate the guarantee value or the quantile of interest. Instead, we can observe the sample sequentially and record only successive minimum or maximum values. Then, measurements are made only at "record-setting" items, and the total number of measurements made is considerably smaller than n, the total sample size. It turns out that we can still estimate the guarantee value from these record-setting measurements. This strategy is extremely useful when items are available for testing, after they are manufactured, but before they are shipped out. Let us say that we have a shipment of wooden beams and we want to make an inference about the breaking strength of these beams. We take a sample of fifty beams to test their breaking strength. In classical sampling, we would destroy all fifty beams. But in the setting of record-breaking data, here is what we would do: We stress the first beam until it breaks and record the breaking stress. The next beam is then stressed only up to the level that broke the first one. If it does not break, we move on to the third beam and stress it up to the value that broke the first beam. If the second beam breaks, its breaking stress is recorded; then we stress the third beam only up to the value that broke the second one. As is obvious, the data will consist of lower and lower breaking stress values. Moreover, the total number of beams broken will surely be less than fifty. Also as mentioned earlier, we will still be able to estimate the required quantile or guarantee value based on the statistical theory of successive minima. Of course, besides estimating the guarantee value, one may want to predict a future record, estimate underlying parameters, or estimate the underlying probability distribution function of the variable being measured. These and other problems have given rise to a plethora of papers and books on record-breaking data.

However, although record values have been around forever, "record-breaking data" as it is called, is relatively new to the field of statistics, owing its birth to Chandler in 1952. Chandler (1952) studied the stochastic behavior of random record values arising from the "classical record model," that is, the record model where the underlying sample from which records are observed is considered to consist of independent identically distributed observations from a continuous probability distribution. Among the many properties that Chandler established for the random record sequence, perhaps the most important and somewhat surprising one was that the expected value of the waiting time between records has infinite expectation. Chandler's work was followed by that of Dwass (1960) and Renyi (1962), who established limit theorems for some of the sequences associated with record-breaking data. Dwass (1964) studied the frequency of records indexed by i, $an \leq i \leq bn$ and showed that this frequency is asymptotically a Poisson count with mean $\ln(b/a)$. Afterward, the subject of record values caught the attention of several mathematicians and statisticians, and work on it increased tremendously. There have been numerous articles on moments of records, characterizations, inference from records, and the like. One only has to survey the recent statistical literature to note the fairly large volume of work that is still being carried out in this field. Also, statisticians have started moving away from the classical model. There are a number of situations where due to improvements in technology or techniques, the underlying population may have a trend in it. Hence the classical record-breaking model will not provide an adequate explanation for these data. In fact, as soon as the basic "independent, identically distributed" assumption for the record model of Chandler is extended to better reflect reality, the problem becomes much harder. A perfect example here is the field of sports. Improved training techniques, diet, health care, and so on, all lead to better performances. Thus the simple record model cannot explain sports records due perhaps to the changing underlying population. Yang (1975), Ballerini and Resnick (1985), and Smith (1988) are just a few of the authors who have moved away from the simple model and have studied models that allow for a changing population.

Although the literature on record values is not enormous compared to other subject areas in statistics, today there are over 300 papers and several books published on record-breaking data. With such a volume of work on records and record-breaking data, it is imperative that related results be brought together in one place. That has been the purpose of most of the books published on record-breaking data and that is also the purpose of this book. Most of the earlier literature on this topic has focused on the stochastic behavior of records, prediction of future record values, and characterization problems. Inference, both parametric and nonparametric, followed later. The manuscripts on record-breaking data have also followed the same trend. With the exception of the book by Arnold et al. (1998), all other books have focused on the stochastic behavior of records, characterizations, and prediction. Arnold et al. (1998) presented a comprehensive review of most of the results on records, including a chapter devoted to the results on inference from record-breaking data. However, that chapter is somewhat brief. The authors focused mainly on estimation of parameters from record-breaking data and not on the general problem of parametric and nonparametric inference from such data. Gulati and Padgett (1994d) gave a brief survey on estimation from such data up until that time.

The purpose of this present monograph is then to fill the gap mentioned above. We focus on cataloging the results on nonparametric inference from record-breaking data. The general problem of parametric and nonparametric inference from record-breaking data has its birth in two articles by Samaniego and Whitaker (1986, 1988). In the first paper (Samaniego and Whitaker, 1986), they develop and study the properties of the maximum likelihood estimator of the mean of an underlying exponential distribution. In their 1988 paper, however, they use record-breaking data to develop a nonparametric maximum likelihood estimator of the underlying distribution function. Under repeated sampling, the nonparametric maximum likelihood estimator is shown to be strongly consistent and asymptotically normal. Their estimator has since been used to develop and study properties of smooth nonparametric function estimates by Gulati and Padgett (1992, 1994, 1995), among others. Besides function estimation from record-breaking data, there are also results on distribution-free tests and nonparametric prediction from such data, as well as a paper on nonparametric Bayesian estimation from record-breaking data (Tiwari and Zalkikar, 1991). All of these results are catalogued here.

In view of the purpose of the monograph, the layout of the book follows. First, the problem of record-breaking data is defined and the notation introduced. This is done in Chapter 2. We also present a summary of the stochastic results on record-breaking data in Chapter 2. Because of the existence of several manuscripts on stochastic results from record-breaking data, and especially in light of the recent monograph by Arnold et al. (1998), the presentation on stochastic results is somewhat brief. In Chapter 3, we discuss some of the major results on parametric inference from such data. Expressions for the estimates of the parameters for various distributions can be found in the book by Arnold et al. (1998). Hence they are not repeated here. Work along the lines of Samaniego and Whitaker (1986), however, has not been discussed by Arnold et al. (1998) and therefore is presented in Chapter 3. The main emphasis of this book, however, begins in Chapter 4. There we tabulate and discuss all the known work on nonparametric inference from such data, starting with the distribution-free tests of Foster and Stuart (1954), leading up to Samaniego and Whitaker's work. Later chapters present some details of the work done in nonparametric function estimation and other results in more recent years. Finally, we consider models that incorporate trend and give a brief outline of some of the work done there.

2
Preliminaries and Early Work

How does one describe record-breaking data in a statistical framework? There are several models for such data, and the classical record model is described here. This model arises, for example, in industrial quality control experiments and destructive stress testing, where one records successive minimum values. As mentioned in Chapter 1, in such experiments one is often interested in estimating a guarantee value or a population quantile. In the classical record model this is done by observing the data sequentially and recording only successive minimum values (since the quantile of interest is normally a lower quantile). Thus, one measures only "record-setting" items and in general, the number of measurements made is considerably smaller than the total sample size. This "measurement savings" is important when the measurement process is costly, time consuming, or destructive.

Consider again the wooden beam example. Suppose a building code prohibits the use of a particular type of beam unless it has probability at least 0.95 of surviving some severe stress, x (see Glick, 1978). In other words, the fifth percentile $x_{0.05}$ should satisfy $x_{0.05} \geq x$. Since it is always better to underestimate the percentile than overestimate it, one considers the smallest failure stress observed in laboratory testing. It is safe to assume that for a large sample, this point will lie below the distribution's fifth percentile. In fact from Glick (1978), the breaking stress of the weakest item in a sample of size 90 lies below the fifth percentile with a probability of 0.99; that is, this minimum value will be the 0.99 tolerance limit for the fifth percentile of the distribution. So we may take a random sample of 90 beams, with our goal being the measurement of the breaking stress of the weakest beam. We want to destroy only a few of the beams and so record sampling is one way to measure the weakest beam. As mentioned in Chapter 1, the breaking stress value of the first beam is our first record value. Thereafter, successive beams are stressed only up to the value at which the previous breakage occurred, with smaller and smaller breaking stress values being recorded until the sample has been exhausted. Note, of course, that the last breaking stress value recorded will be the breaking stress of the weakest item and the estimate of the required tolerance limit. Moreover, on average, we will destroy only about 5 beams (Glick, 1978), leaving the remainder intact for shipment.

The verbal description above is now quantified in a more exact mathematical framework. Then some stochastic properties and statistical characterizations are briefly summarized in the remainder of this chapter.

2.1 Notation and Terminology

The notation and basic statistical framework for the record values (successive minima) is now introduced. Let Y_1, Y_2, . . . be a random sample from a continuous cumulative distribution function (c.d.f.) F with density function f. Then, since only successive minimum values are recorded, the observed data consist of the sequence X_1, K_1, X_2, K_2, . . . , X_r, K_r, where $X_1 = Y_1$, X_i , $i = 2, 3, . . . , r$, is the ith new minimum, and K_i is the number of trials following the observation of X_i to obtain a new record (or to exhaust all available observations in the case of $i = r$, K_r). The sampling schemes for generating these data are:

1) Data are obtained via the *inverse sampling scheme*, where items are presented sequentially and sampling is terminated when the rth record is observed. In this case, the total number of items sampled N_r is a random variable and K_r is defined to be 1 for convenience.

2) Records are obtained under the *random sampling scheme*, that is, a random sample, Y_1, Y_2, . . . , Y_n, from c.d.f. F is examined sequentially and successive minimum values are recorded. For this sampling scheme the number of records R_n obtained is a random variable and, given a value of r, $\sum_{i=1}^{r} K_i = n - 1$.

To understand the terminology better, we look at another example. Consider the process of measuring the thickness of a manufactured item using a micrometer. In order to measure the minimum thickness of $n = 100$ items one first measures an item at random. The thickness of this item then is $X_1 = Y_1$. The gap in the micrometer created by the first item serves as a standard in judging subsequent items, and a new measurement is made only if a subsequent item fits inside this gap. Hence, if the second measurement is made at the sixth trial, then $X_2 = Y_6$, $K_1 = 5$, and $N_2 = 6$ (note that N_1 is always equal to one). Now the gap created by this sixth item serves as a standard for judging subsequent items. Once again, by using this method, the number of actual measurements made will be substantially less than 100, and yet it will serve equally well in determining the minimum thickness.

Regardless of the sampling scheme, we define the following sequences: $\{X_i, 1 \leq i \leq r\}$ is the *record value sequence*, $\{N_i, 1 \leq i \leq r\}$ is defined to be the *record time sequence* (note that $N_1 = 1$ by default), and finally, $\{K_i, 1 \leq i \leq r\}$ is the *interrecord time sequence*. With the above notation and terminology, we have what is called the *Classical Record Model*.

2.2 Stochastic Behavior

The stochastic behavior of the classical record model was first studied by Chandler (1952) who showed that the record times (N_is) and the interrecord times (K_is) both had infinite expectation, although the mode of the K_is was one. Chandler also gave an expression for the joint distribution of $X_1, X_2, \ldots, X_r$, and obtained tables for the percentile points for X_i for $i = 1, 2, \ldots, 9$, for the normal and the rectangular distributions.

The fact that N_2 has an infinite expectation discouraged many statisticians from working on record-breaking data for a while (see Galambos, 1978). Development began again with Dwass (1960) and Renyi (1962) who gave "strong law of large numbers"- and "central limit theorem"-type results for the R_ns and the N_is. Dwass (1964) also showed the frequency of the record highs among the observations indexed by i, $an \leq i \leq bn$, where n is the sample size and $0 < a < b$ is asymptotically a Poisson count with mean $\ln(b/a)$. Since Dwass, a number of statisticians have investigated the behavior of record values. (See Neuts, 1967, Resnick, 1973(a,b,c), and Shorrock, 1972(a,b), 1973, for some of the articles on the subject.) Glick (1978) gave an informal summary of results to that date. Review articles have also been written by Galambos (1978), Nagaraja (1988), and Nevzorov (1987). In addition, the topic of records and record-breaking data has been discussed in a number of books. Besides the very thorough and comprehensive work by Arnold et al. (1998), work on records has been reviewed in the books by Galambos (1987), Resnick (1987), and Ahsanullah (1995). What follows next is a brief summary of the stochastic results about the number of records R_n, record values the X_is, the interrecord times K_is, and the record times N_is. For details of these results, the reader is referred to the book by Arnold et al. (1998).

Record Values

a)   The joint distribution of $X_1, X_2, \ldots, X_r$ is given by the probability density function (Chandler, 1952, and Glick, 1978)

$$g(x_1, x_2, \ldots, x_r) = f(x_r) \prod_{i=1}^{r-1} \frac{f(x_i)}{1 - F(x_i)}.$$

b) Define $H(y) = -\ln (1 - F(y))$. Then we have the large sample results that (Resnick, 1973a)

   i) $[H(x_r) - r]/\sqrt{r}$ is asymptotically normal with mean 0 and variance 1, and

   ii) $H(x_r)/\sqrt{r}$ converges to one with probability one.

In fact, Resnick (1973a) characterized the three types of limit distributions to which the record values X_r can converge as $r \to \infty$. Depending on the underlying distribution F, a record value sequence satisfies exactly one of the following convergences in distribution as $r \to \infty$.

i) $P\left\{ \dfrac{X_{N_r} - G^{-1}(r)}{G^{-1}(r + \sqrt{r}) - G^{-1}(r)} < x \right\} \to \Phi(x),$

ii) $P\left\{ \dfrac{X_{N_r}}{G^{-1}(r)} < x \right\} \to \begin{cases} 0, & x < 0 \\ \Phi(\alpha \ln x)), & x \geq 0 \end{cases}$

and

iii) $P\left\{ \dfrac{X_{N_r} - \bar{x}}{\bar{x} - G^{-1}(r)} < x \right\} \to \begin{cases} \Phi(-\alpha \ln(-x)), & x < 0 \\ 1, & x \geq 0, \end{cases}$

where $\Phi(x)$ denotes the standard normal distribution function.

Frequency of Records

Results for the statistical behavior of the number of records R_n in a random sample of size n are listed next:

   a) $E(R_n) = 1 + \dfrac{1}{2} + \dfrac{1}{3} + \cdots + \dfrac{1}{n}$ (Glick, 1978);

   b) $\mathrm{Var}(R_n) = \displaystyle\sum_{i=1}^{n} \frac{1}{i} - \sum_{i=1}^{n} \frac{1}{i^2}$ (Glick, 1978).

   c) $[\ln(R_n) - n]/n^{1/2}$ is asymptotically normal with mean zero and variance one (Resnick, 1973c).

   d) $R_n/\ln(n)$ converges to one with probability one (or almost surely, a.s.) as the sample size $n \to \infty$ (Renyi, 1962). That is, as n increases, about $\ln(n)$ records will be found in a random sample of n values.

e) The frequency of record highs among the observations indexed by i, $an \leq i \leq bn$ ($0 < a < b$) is asymptotically a Poisson count with mean $\ln(b/a)$ (Dwass, 1964).

Record Times

The record times N_i also have interesting stochastic properties, as described next.

a) The value of N_i does not depend on the underlying distribution function (Chandler, 1952).

b) $[\ln(N_r) - r]/\sqrt{r}$ is asymptotically normal with mean 0 and variance 1 (Renyi, 1962, Resnick, 1973c).

c) $N_r / \ln(r)$ converges to one with probability one as the number of records $r \to \infty$ (Renyi, 1962, Dwass, 1960, and Galambos, 1978).

d) The distribution of the ratio N_r / N_{r+1} is asymptotically uniform over the unit interval (Tata, 1969).

e) The successive ratios N_r / N_{r+1}, N_{r+1} / N_{r+2}, . . . are asymptotically independent uniform variates (Shorrock, 1972b and Resnick, 1973c).

Waiting Time Between Records

For the interrecord time sequence, perhaps the most surprising result is the infinite expected value, the first property below. The asymptotic distributional behavior of K_r is similar to that of N_r.

a) $E(K_i) = \infty$ for all i, although the mode of the K_is is 1 (Chandler, 1952).

b) $[\ln(K_r) - r]/\sqrt{r}$ is asymptotically normal with mean 0 and variance 1 (Neuts, 1967).

c) $K_r / \ln(r)$ converges to one with probability one as $r \to \infty$ (Neuts, 1967).

In addition to the brief summary presented above of some of the main results on records, it can be shown that the record time sequence, the record-value sequence, and the sequence of the number of records each forms a Markov chain. There are also results on the moments of the record value sequence (both parametric and nonparametric) and on characterizations. Once again, the reader is referred to the book by Arnold et al. (1998) for a detailed discussion on all of these results.

3
Parametric Inference

3.1 General Overview

The roots of inference from record-breaking data lie in Foster and Stuart's (1954) research. They developed distribution-free tests based on record values to determine if upper records were from an i.i.d sequence of observations (details of the test are presented in the next chapter). After that, however, statistical inference from record-breaking data remained virtually unexplored until the late 1970s when some statisticians started investigating the important and interesting problem of predicting future records.

During the 1980s, there was a considerable amount of work done in the area of prediction of records. Assuming that F was a two-parameter exponential distribution function, Ahsanullah (1980) obtained the best linear unbiased predictor and the best linear invariant predictor of a future record X_m, based on observed record values X_i, $1 \le i \le n$, for $n < m$, using standard least squares theory. Dunsmore (1983) provided tolerance regions as well as Bayesian predictive distributions for $X_m - X_n$ for record-breaking data from an exponential distribution. Predictors of future records from three extreme value distributions were also studied by Weissman (1978) and Nagaraja (1988). For the uniform, normal, and extreme value distributions, Tryfos and Blackmore (1985) obtained minimum-variance linear unbiased estimators of the parameters and best linear unbiased predictors of the future records using generalized least squares. Samaniego and Kaiser (1978) obtained estimates of the upper bound of a uniform distribution function based on the X_is. In all these papers, the authors worked with upper record values, that is, the case where successive maxima were recorded.

Estimation of the parameters of the underlying distribution from which record values are observed is another important problem. Work was done in this direction by Tryfos and Blackmore (1985) as mentioned earlier. Balakrishnan and Cohen (1991) developed the best linear unbiased estimator (BLUE) of the scale parameter σ, when the observed data consist of upper records from a scale parameter distribution with c.d.f. $F(x/\sigma)$. Balakrishnan and Chan (1994) developed BLUEs for the underlying parameters of the one-parameter Rayleigh and the one-parameter Weibull distributions. This work was extended to the logistic distribution by Balakrishnan et al. (1995) and to the normal distribution by Balakrishnan and Chan (1998).

The results mentioned above are only the tip of the iceberg in the area of inference. There are a number of other articles and results on parametric inference from record-breaking data, including interval estimation and hypothesis testing. We do not present these results here, since they are easily found in Nagaraja (1988) or Arnold et al. (1998). The purpose of this chapter is to elaborate on some of the results on parametric inference not touched upon in the book by Arnold et al. (1998). Most such results are related to the papers by Samaniego and Whitaker (1986, 1988), and they are presented in the next section.

3.2 Parametric Inference Due to Samaniego and Whitaker

The general problem of classical parametric and nonparametric inference from record-breaking data was first addressed by Samaniego and Whitaker (1986, 1988). Assuming that the sample consisted of lower record values (successive minima) from an exponential distribution with mean μ, Samaniego and Whitaker (1986) obtained and studied properties of the maximum likelihood estimator (MLE) of μ under both the inverse sampling scheme and the random sampling scheme. This was followed in 1988 by the paper on nonparametric maximum likelihood estimation of the underlying survival function from record-breaking data. Here, we start with parametric inference from inversely sampled records.

3.2.1 Exponential Distribution—Inverse Sampling

Suppose that $X_1, K_1, X_2, K_2, \ldots, X_r, K_r$ is the observed record sample from an underlying exponential distribution with mean μ. Hence, the underlying density function is given by

$$f(x) = \lambda \exp(-\lambda x), \ x > 0, \ \text{with} \ \mu = \frac{1}{\lambda} \ . \tag{3.2.1.1}$$

Under inverse sampling, the conditional density of X_i, given $X_1, K_1, X_2, K_2, \ldots, X_{i-1}, K_{i-1}$ is

$$f(x \mid X_{i-1} = x_{i-1}) = \frac{f(x)}{F(x_{i-1})} I_{(0, \, x_{i-1})}(x), \tag{3.2.1.2}$$

where I_A is the indicator function of the set A, $I_A(x) = 1$ if x is in A and 0 if not. As is clear by the nature of the data, the above density depends on $X_1, K_1, \ldots, X_{i-1}, K_{i-1}$ only through X_{i-1}. Similarly, the density of K_i given $X_1, K_1, \ldots, X_{i-1}, K_{i-1}, X_i$ depends only on the value of X_i and is the geometric distribution with probability mass function given by

$$P(K_i = k \mid X_i = x_i) = F(x_i)(1 - F(x_i))^{k-1}, \quad k = 1, 2, 3, \ldots. \tag{3.2.1.3}$$

It follows then that the likelihood function associated with the sequence $\{X_1, K_1, \ldots, X_r, K_r\}$ is

$$L = \prod_{i=1}^{r} f(x_i)(1 - F(x_i))^{k_i - 1} I_{(0, x_{i-1})}(x), \tag{3.2.1.4}$$

which for the density in (3.2.1.1) reduces to

$$L = \lambda^r \exp\left[-\left(\lambda \sum_{i=1}^{r} k_i x_i\right)\right] \prod_{i=1}^{r} I_{(0, x_{i-1})}(x_i).$$

The derivation of the MLE of λ and therefore of μ is almost immediate. The MLE is given by

$$\hat{\mu} = \frac{\sum_{i=1}^{r} K_i X_i}{r}, \tag{3.2.1.5}$$

where $K_r = 1$.

In investigating the distributional properties of (3.2.15), Samaniego and Whitaker (1986) consider the moment generating function $m(t)$ of $\sum_{i=1}^{r} K_i X_i$. The m.g.f is derived easily using the likelihood function L as follows.

$$m(t) = E(\exp(t \sum_{i=1}^{r} k_i x_i)) =$$

$$= \sum_{k, x} \left[e^{\left\{ t \sum_1^r k_i x_i \right\}} \right] \lambda^r e^{\lambda \sum_1^r k_i x_i} \, dx$$

$$= \frac{\lambda^r}{(\lambda - t)^r} \sum_{k,x} (\lambda - t)^r e^{-(\lambda - t)\sum_i k_i x_i} \, dx$$

$$= (1 - \mu t)^{-r} \quad \text{for } t < \mu^{-1}, \tag{3.2.1.6}$$

where the notation

$$\sum_{k,x} g(x_1, k_1, \ldots, x_r, k_r) \, dx$$

means the integration over all x_is and summation over all k_is of $g(\cdot)$ as explained by Samaniego and Whitaker (1986). Note that (3.2.1.6) is the moment-generating function for the gamma distribution with shape parameter r and scale parameter μ. The unbiasedness of $\hat{\mu}$ then follows easily, since

$$E(\hat{\mu}) = \frac{E\left(\sum_{i=1}^{r} K_i X_i\right)}{r} = \frac{r\mu}{r} = \mu.$$

Also, note that since $\hat{\mu}$ is a function of a complete sufficient statistic, $\hat{\mu}$ is the minimum variance unbiased estimator (MVUE) of μ with the variance of $\hat{\mu}$ given by

$$\text{Var}(\hat{\mu}) = \frac{\mu^2}{r}.$$

The consistency of $\hat{\mu}$ (as $r \to \infty$) follows from Chebsyhev's inequality, and since $\sum_{i=1}^{r} K_i X_i$ is distributed as the sum of i.i.d. exponential random variables, the asymptotic normality of $\hat{\mu}$ follows from the Central Limit Theorem.

3.2.2 Exponential Distribution—Random Sampling

In the case of random sampling, the variable R, the number of records in a random sample of size n, is a random variable. Now, the likelihood function of the variables $R, X_1, K_1, X_2, K_2, \ldots, X_R, K_R$ is

$$L = \left\{ \prod_{i=1}^{r-1} p(x_i | x_{i-1}) \, p(k_i | x_i) \right\} p(x_r | x_{r-1}) \, P(R = r | x_1, k_1, \ldots, x_{r-1}, k_{r-1}, x_r)$$
$$\times P(k_r | x_1, k_1, \ldots, k_{r-1}, R = r)$$

$$= \left\{ \prod_{i=1}^{r-1} \frac{f(x_i)}{F(x_{i-1})} F(x_i)(1-F(x_i))^{k_i-1} \right\} \left\{ \frac{f(x_r)}{F(x_r)} (1-F(x_r))^{k_r-1} \times 1 \right\}$$

$$= \prod_{i=1}^{r} f(x_i)(1-F(x_i))^{k_i-1} . \qquad (3.2.2.1)$$

Note that the likelihood equation in (3.2.2.1) is identical to the one in (3.2.1.4), so that we have identical likelihoods for the exponential distribution for both sampling schemes. Thus, for the random sampling scheme, the maximum likelihood estimator of the mean μ is

$$\tilde{\mu} = \frac{\sum_{i=1}^{R} K_i X_i}{R} . \qquad (3.2.2.2)$$

Even though the structural form of the estimator is identical under both sampling schemes, the distributional theory and statistical characteristics of the two estimators are quite different. Here, the conditional distribution of $\sum_{i=1}^{R} K_i X_i$ given $R = r$ is identical to that of a gamma random variable with shape parameter r and scale parameter μ, but the unconditional distribution of $\sum_{i=1}^{R} K_i X_i$ is that of a mixture of gamma distributions. Specifically, we write it as

$$\sum_{i=1}^{R} K_i X_i \sim \sum_{r=1}^{n} P_n(R = r) \Gamma(r, \mu) \qquad (3.2.2.3)$$

where $\Gamma(r, \mu)$ represents a gamma random variable with shape r and scale μ. Hence, conditional on the value of R, $\tilde{\mu}$ is a minimum variance unbiased estimator of μ, but in fact, Samaniego and Whitaker (1986) show that the unconditional estimator is not the unique MVUE (UMVUE) of μ. This is done by constructing an estimator for μ that is unbiased and has a smaller variance than $\tilde{\mu}$. Asymptotic normality has also not been established for the unconditional estimator. Before examining why $\tilde{\mu}$ is not the UMVUE of μ, we look at an example illustrating its calculation.

Example 3.2.2.1:
Samaniego and Whitaker used the above methods to calculate the underlying mean for the successive failure times for air conditioners in Boeing 747 airplanes. These data were first introduced by Proschan (1963) and since then have been studied by a number of authors. Proschan (1963) tested and accepted the hypothesis that the successive

Table 3.2.1. Successive Minima, Plane 7914

i	X_i	K_i
1	50	1
2	44	3
3	22	2
4	3	18

failure times of the air conditioner units were i.i.d. exponential. Samaniego and Whitaker used the data for Plane 7914 which consisted of $n = 24$ planes. Successive minima obtained from this plane are shown in Table 3.2.1. For these data, the estimated mean life is

$$\tilde{\mu} = (50 + (44 \times 3) + (22 \times 2) + (3 \times 18))/4 = 70.$$

Note that had one calculated the straight average, using all 24 failure times, one would have estimated the mean of the sample as 64.125.

Returning now to the theoretical behavior of $\tilde{\mu}$, we examine the construction of an estimator of μ which is unbiased and has a smaller variance than $\tilde{\mu}$. Note that if we let $S = \sum_{i=1}^{R} K_i X_i$, then in fact, $\tilde{\mu} = S/R$. To show that $\tilde{\mu}$ is not the UMVUE of μ, Samaniego and Whitaker (1986) construct an estimator of the form $\tilde{\mu}_a = a(R)S$ which is unbiased and has a smaller variance than $\tilde{\mu}$. Note that for $\tilde{\mu}_a$ to be unbiased, we must have that $E(\tilde{\mu}_a) = E(E(\tilde{\mu}_a \mid R = r)) = \sum_{r=1}^{n} a(r) r \mu \, p_n(r) = \mu$, where $p_n(r) = P_n(R = r)$. That is, one must have $\mathbf{p}^T\mathbf{a} = 1$, where $\mathbf{p}^T$ and $\mathbf{a}^T$ are n-dimensional row vectors with elements $p_i = ip_n(i)$ and $a_i = a(i)$, $i = 1, 2, \ldots, n$, respectively.

Now, if $\tilde{\mu}_a$ is unbiased, then Var $(\tilde{\mu}_a) = E(\tilde{\mu}_a^2) - \mu^2 = \mu^2 \{\mathbf{a}^T\mathbf{P}\mathbf{a} - 1\}$, where P is an $n \times n$ diagonal matrix with elements $P_{ii} = i(i+1)p_n(i)$. So the idea now is to find the vector **a** that minimizes $\mathbf{a}^T\mathbf{P}\mathbf{a}$ subject to the constraint $\mathbf{a}^T\mathbf{P} = 1$. This is done easily by using Lagrange multipliers and gives the following estimator for μ,

$$\tilde{\mu}_a \text{ for which } a^*(i) = \frac{[E(R/R+1)]^{-1}}{(i+1)}. \qquad (3.2.2.4)$$

The variance of $\tilde{\mu}_a$ can be shown to be

$$\text{Var}(\tilde{\mu}_a) = \mu^2 \frac{E(1/R+1)}{E(R/R+1)}. \tag{3.2.2.5}$$

The expected values in the above equations can be calculated from the distribution of R. Although the estimator $\tilde{\mu}_a$ is unbiased conditionally and has a smaller variance than the MLE, it is, however, not unbiased unconditionally. In fact, $\tilde{\mu}_a$ has a negative bias for small values of R and a positive bias for large values of R.

3.3 Parametric Inference Beyond Samaniego and Whitaker

Samaniego and Whitaker's (1986) work was extended by Hoinkes and Padgett (1994) to the two-parameter Weibull distribution. They considered maximum likelihood estimation of the parameters for lower record values obtained from the Weibull distribution under the random sampling case and studied their properties via computer simulations. Assuming that the record-breaking data are obtained from the density function $f(y) = \lambda\beta y^{\beta-1} \exp(-\lambda y^{\beta})$, $y > 0$, the likelihood function is given by

$$L = \prod_{i=1}^{R} \left\{ \lambda\beta x_i^{\beta-1} \exp[-\lambda x_i^{\beta}] \, (1 - (1 - \exp[-\lambda x_i^{\beta}]))^{k_i-1} \right\}$$

$$= \prod_{i=1}^{R} \left\{ \lambda\beta x_i^{\beta-1} \exp[-\lambda x_i^{\beta}(k_i)] \right\}. \tag{3.3.1}$$

Taking logarithms and then differentiating with respect to λ and β leads to the following equations which must be solved to obtain the maximum likelihood estimates of the parameters λ and β,

$$h(\beta) = r/\beta + \sum_{i=1}^{r} \ln(X_i) - (r/\sum_{i=1}^{r} X_i^{\beta} K_i) \sum_{i=1}^{r} (X_i^{\beta} K_i \ln(X_i)) = 0, \tag{3.3.2}$$

where λ is expressed in terms of β,

$$\lambda = \frac{r}{\sum_{i=1}^{r} X_i^{\beta} K_i}.$$

The system of equations above cannot be solved explicitly and the MLEs must be found by iterative methods. The mean-squared errors (MSE) of these estimators were studied via simulations for different sample sizes and various values of β. The value of the scale parameter, λ, was held constant at one. Hoinkes and Padgett (1994) found that in general, the MLE of λ was very inefficient (MSEs were in the 10^{31} to 10^{36} range). They also found that the MLE was in fact inconsistent, since the mean-squared error of the estimator increased as n gets large for small values of β. However, as the value of β increased, the MSE of the estimator of λ decreased. This was explained by the fact that as β increased, the size of the record sample increased leading to more information and therefore better estimators. In an attempt to get the estimator of the scale parameter to behave a little better, a reparameterization of the Weibull distribution was explored with scale parameter given by $\alpha = \lambda^{-1/\beta}$. Although the MLEs of α for the reparameterized distribution did behave better, the final conclusion of the paper is that maximum likelihood estimation of the distribution function for the Weibull distribution from record-breaking data was not efficient. The characteristics of the record-breaking data from the Weibull were not consistently reflective of the true population. Could this perhaps be explained by the fact that the minimum of i.i.d. Weibull random variables is again a Weibull random variable with a different scale parameter? Table 3.3.1 presents some results from their simulations.

Table 3.3.1. Mean-Squared Error Results from the Simulations

β	n	Complete Sample MSE		Record Sample MSE		Relative Efficiencies	
a) $\lambda = 1$		λ	β	λ	β	λ	β
	100	0.01257	0.00440	1.26E32	0.52588	9.71E-31	0.00837
0.8	500	0.00212	0.00085	9.97E32	0.11507	2.14E-31	0.00738
	1000	0.00105	0.00038	3.83E36	0.10913	2.75E-36	0.00349
	100	0.01261	0.00686	1.29E32	0.70976	9.74E-31	0.00971
1.0	500	0.00213	0.00113	9.97E32	0.17980	2.14E-32	0.00738
	1000	0.00105	0.00059	3.83E36	0.17052	2.75E-36	0.00349
	100	0.01257	0.10767	2.66E5	2.5023	4.73E-4	0.04303
4.0	500	0.00216	0.02121	2.09E5	1.2775	6.99E-5	0.01660

Table 3.3.1. (Continued)

β	n	Complete Sample MSE		Record Sample MSE		Relative Efficiencies	
		α	β	α	β	α	β
b) $\alpha = 1$							
	100	0.01821	0.00440	0.90731	0.52588	0.02007	0.00837
0.8	500	0.00328	0.00085	17.993	0.11507	0.00018	0.00737
	1000	0.00164	0.00038	12.213	0.10913	0.00135	0.00935
	100	0.01167	0.00689	0.47652	0.70976	0.02449	0.00970
1.0	500	0.00210	0.00133	2.7884	0.17980	0.00075	0.00737
	1000	0.00105	0.00059	2.4663	0.17052	0.00043	0.00349
	100	0.00074	0.10767	0.03041	2.5023	0.02427	0.02205
4.0	500	0.00013	0.02121	0.03215	1.2775	0.00414	0.01660
	1000	6.56E-5	0.00957	0.03464	0.95216	0.00189	0.01005

The poor performance of the MLEs for the Weibull led to a parametric comparison of the random sampling schemes and the inverse sampling schemes. During simulations, Hoinkes and Padgett found that sometimes for samples as large as 1000, one obtained only two or three records. It appeared then perhaps that estimates obtained from inverse sampling (where one can fix the number of records observed) might perform better. Hence, Berger and Gulati (2000) compared the MLEs for the random and inverse sampling case for some distributions. The mean-squared errors and biases of the estimates were computed and compared via simulations for the exponential, Weibull, Pareto, and the Type I extreme value (or the Gumbel) distributions. It was found that there was no significant difference in their performance for the two sampling schemes for the exponential and the extreme value distributions. However, for the Weibull and the Pareto distributions, the inverse sampling scheme was found to yield much better estimates than the random sampling scheme. As an example, for the Weibull distribution the MSE for $\hat{\lambda}$ (the MLE of λ) ranged from 10^{10} to 10^{146}, depending on the sample size, and was not very stable. For the inverse sampling scheme, the MSE of the MLE $\hat{\lambda}$, ranged from the order of 10^{4}

to 10^{90} for comparable values of r. Moreover, the estimates were very consistent. Hence although neither of the schemes yielded efficient estimators for λ, the inverse sampling scheme did yield better estimates than the random sampling scheme. A similar phenomenon was observed for the Pareto distribution with inverse sampling yielding much better estimates than the random sampling scheme. Berger and Gulati's article ended by concluding that perhaps as long as distributions are "well-behaved," there is no significant difference between the two sampling schemes. But, for badly behaved distributions (and therefore estimators) inverse sampling yields better estimates. Moreover, inverse sampling was also found to be more efficient in terms of the total number of items examined. Although a random sample of size $n = 1000$ yields about 7.5 records on average, the average sample size to yield 7 records for inverse sampling was only about 258 in their simulation studies. Tables 3.3.2 and 3.3.3 give the mean-squared errors for the Weibull distribution for the two schemes.

Table 3.3.2. Random Sampling, Weibull Distribution, $\lambda = 1$

		Sample Size				
		31	69	128	245	
β	0.8	0.397×10^{96}	0.947×10^{131}	0.596×10^{26}	0.258×10^{146}	$\hat{\lambda}$ MSE
		3.122	3.269	0.212	1.325	$\hat{\beta}$ MSE
	1.0	0.397×10^{96}	0.947×10^{131}	0.596×10^{26}	0.258×10^{146}	$\hat{\lambda}$ MSE
		4.879	5.108	0.332	2.071	$\hat{\beta}$ MSE
	2.0	0.269×10^{49}	0.135×10^{45}	0.596×10^{26}	0.909×10^{65}	$\hat{\lambda}$ MSE
		11.666	6.678	1.327	3.363	$\hat{\beta}$ MSE
	4.0	0.769×10^{29}	0.135×10^{45}	0.596×10^{26}	0.597×10^{39}	$\hat{\lambda}$ MSE
		24.348	17.819	5.307	6.818	$\hat{\beta}$ MSE
	12.0			0.615×10^{12}	0.165×10^{10}	$\hat{\lambda}$ MSE
				31.936	20.750	$\hat{\beta}$ MSE

Table 3.3.3. Inverse Sampling, Weibull Distribution, $\lambda = 1$

		No. of Records				
		3	4	5	7	
	0.8	0.423×10^{90}	0.6242×10^{20}	0.111×10^{9}	0.976×10^{4}	$\hat{\lambda}$ MSE
		7.559	1.031	0.297	0.078	$\hat{\beta}$ MSE
	1.0	0.407×10^{83}	0.6242×10^{20}	0.111×10^{9}	0.976×10^{4}	$\hat{\lambda}$ MSE
β		9.335	1.611	0.465	0.121	$\hat{\beta}$ MSE
	2.0	0.962×10^{63}	0.6242×10^{20}	0.111×10^{9}	0.976×10^{4}	$\hat{\lambda}$ MSE
		25.941	6.443	1.859	0.485	$\hat{\beta}$ MSE
	4.0	0.363×10^{36}	0.6242×10^{20}	0.111×10^{9}	0.976×10^{4}	$\hat{\lambda}$ MSE
		66.614	21.320	7.437	1.938	$\hat{\beta}$ MSE
	12.0			0.330×10^{7}	0.976×10^{4}	$\hat{\lambda}$ MSE
				46.910	17.449	$\hat{\beta}$ MSE

Maximum likelihood estimation from univariate distributions was also studied by Qasem (1996). For a sequence of upper records, Qasem (1996) developed MLEs for θ for the uniform distribution on (0, *θ*), the exponential distribution with mean *θ*, and the Pareto distribution with shape parameter *θ*.

To develop the maximum likelihood estimators, Qasem (1996) used the well-known results given next about the joint density and the marginal density of record values.

Suppose that we have m upper record values (that is the case where successive maxima are recorded), $x(1)$, $x(2)$, . . . , $x(m)$, from a continuous distribution with cumulative distribution function F, and probability density function f. In addition, assume as before that the records are drawn from a random sample of size n denoted by $\{Y_i, 1 \leq i \leq n\}$. Then the joint density of the record values and the marginal density of the jth record values are given by

$$g_{X(1), X(2), \ldots, X(m)}(x(1), \ldots, x(m)) = f(x(m)) \prod_{i=1}^{m-1} \frac{f(x(i))}{1 - F(x(i))} \quad (3.3.3)$$

and

$$g_{X(j)}(x) = \frac{\left[-\ln(1-F(x))\right]^{j-1}}{(j-1)!} f(x).$$
(3.3.4)

Applying (3.3.3) and (3.3.4) to the uniform, exponential, and Pareto distributions gives us the following.

a) The uniform distribution on $(0, \theta)$.

From (3.3.3), the likelihood function of $x(1), x(2), \ldots, x(m)$ for an underlying uniform distribution is given by

$$L\ (x(1), x(2), \ldots, x(m);\ \theta) = \frac{1}{\theta} \prod_{i=1}^{m-1}\left(\frac{1}{\theta - x(i)}\right),\quad \theta > x(m).$$

Note that the above function attains its maximum when θ is minimum, that is, when $\tilde{\theta} = X(m)$, the highest record value.

From (3.3.4), the probability density function of $X(m)$ is

$$g_{X(m)}(x) = \frac{\left[-\ln(1-\frac{x}{\theta})\right]^{m-1}}{(m-1)!} \frac{1}{\theta} I_{(0,\theta)}(x),$$

where $I_{(0,\theta)}(\cdot)$ is the indicator function on $(0, \theta)$.

The mean and variance of $\tilde{\theta}$ are thus easily calculated as

$$E(\tilde{\theta}) = E(X(m)) = \theta\,(1 - 2^{-m}) \quad \text{and} \quad Var(\tilde{\theta}) = \theta^2\,[3^{-m} - 4^{-m}].$$

It is easy to see then that $\tilde{\theta}$ is asymptotically unbiased and consistent for θ. Moreover, note that the complete sample MLE for the uniform distribution is the nth order statistic $Y_{(n)}$, with mean and variance given by

$$E(Y_{(n)}) = \frac{n}{n+1}\theta \quad \text{and} \quad Var(Y_{(n)}) = \frac{n}{(n+2)(n+1)}\theta^2.$$

Therefore, the relative efficiency of the MLE obtained via the entire random sample versus the record sample is

$$r.e. = \frac{n(12)^m}{(n+2)(n+1)^2(4^m - 3^m)}.$$

The results for the exponential and the Pareto distribution are obtained similarly.

b) The exponential distribution with mean θ.
   Here the likelihood function is

$$L(x(1), x(2), \ldots, x(m); \theta)$$
$$= \frac{1}{\theta} \exp(-x(m)/\theta) \prod_{i=1}^{m-1} \frac{\theta^{-1} \exp(-x(i)/\theta)}{\exp(-x(i)/\theta)}$$
$$= \theta^m \exp(-x(m)/\theta).$$

Therefore, the MLE of θ is given by

$$\tilde{\theta} = \frac{x(m)}{m}.$$

Again, as with the uniform distribution, the mean and the variance of the estimator can be easily calculated and the MLE shown to be unbiased and consistent. The relative efficiency of the entire random sample (size n) MLE versus the record sample (size m) MLE is m/n.

c) The Pareto distribution defined on $(1, \infty)$ with shape parameter θ.
   For this distribution, the underlying density function is

$$f(y) = \theta x^{-(\theta+1)}, x > 1.$$

Therefore, the likelihood function of the record values is calculated as

$$L(x(1), x(2), \ldots, x(m); \theta) = \theta^m [x(m)]^{-\theta} \prod_{i=1}^{m} x(i).$$

Maximizing the above gives the MLE of θ as

$$\tilde{\theta} = \frac{m}{\ln(x(m))}.$$

Using (3.3.4), note that the marginal density of $x(m)$ is

$$f_{(x(m))}(x) = \frac{[\theta \ln(x)]^{m-1}}{(m-1)!} \theta x^{-(\theta+1)}.$$

Table 3.3.4: Relative Efficiencies: Complete Sample MLE Efficiencies
vs. Record Sample Efficiencies

n	m	Relative Efficiencies		
		Uniform	Exponential	Pareto
20	3	0.09627	0.15	0.0273
50	4	0.04380	0.0800	0.0244
150	5	0.01379	0.0333	0.0131
400	6	0.00548	0.015	0.0070
1000	7	0.00251	0.0070	0.0037

A simple transformation of variables then shows that $\ln(x(m))$ has the gamma distribution with shape parameter m and scale parameter θ, so that

$$E(\tilde{\theta}) = \frac{m}{m-1}\,\theta \text{ and } Var(\tilde{\theta}) = \frac{1}{m-2}\left(\frac{m}{m-1}\right)^2 \theta^2.$$

Once again, it is easy to verify that the MLE obtained from the record sample is asymptotically unbiased. Also, the estimator is consistent. The relative efficiency of the entire random sample (size n) MLE versus the record sample (size m) MLE is given by

$$\frac{m-2}{n-2}\left[\left(\frac{n}{n-1}\right)\left(\frac{m-1}{m}\right)\right]^2.$$

For illustrative purposes, Qasem (1996) calculated the relative efficiencies of the complete-sample MLE versus the record-sample MLE for various values of n (the value of m was fixed at the integer closest to $\ln(n)$). Some of these efficiencies as well as some additional ones are presented in Table 3.3.4.

3.4   The Problem of Prediction

Predicting the future record value—a problem that interests all of us! As an example, consider the following questions that we are always asking ourselves. What will be the *next* record for the 100 m sprint? What will be the *next* highest temperature ever recorded? Will Brazil win the next World Cup and break its own record of being the country with the highest number of World Cup wins? To answer these kinds of questions with any degree of certainty will require the use of past data —in other words, statistical expertise.

Point and interval prediction of future records using past data has dominated scientific research from record-breaking data. First considered by Ahsanullah (1980), the problem of prediction of future records has been studied by several statisticians (see Dunsmore, 1983, Weissman, 1978, Nagaraja, 1984, Ahsanullah, 1993, and Balakrishnan and Chan, 1994, among others). A detailed description of most of the above work has been provided in Arnold et al. (1998). In this section then, we detail some of the same results as well some of the work on prediction not found in Arnold et al. (1998).

Throughout the subsection, as in Qasem (1996), we let $x(0)$, $x(1)$, . . . , $x(m)$ denote the first $(m + 1)$ upper record values observed from a continuous distribution with cumulative distribution function F and density function f. The problem of interest then is to predict the sth record value for some $s > m$. In terms of point prediction, most papers published on this problem have focused on linear predictors of future record values based on past data, in particular, best linear unbiased predictors and best linear invariant predictors. We start with the results on best linear unbiased predictors, or BLUPS.

3.4.1. Best Linear Unbiased Prediction

Suppose F is a location-scale family of distributions with the location parameter represented by μ and the scale parameter by σ. Using the results of Goldberger (1962) for a general linear model, it is easily shown that the BLUP, $x^*(s)$, of the future record value $x(s)$ is given by

$$x^*(s) = \mu^* + \alpha(s)\sigma^* + \omega^T \Sigma^{-1}(\mathbf{X} - \mu^*\mathbf{1} - \sigma^*\alpha), \qquad (3.4.1.1)$$

where μ^*, σ^* are the best linear unbiased estimators of μ and σ, $\mathbf{X}$ is the vector of observed record values, $\mathbf{1}$ is a vector of ones, α is the vector of means of the record values from the standard distribution, $\Sigma = (\sigma_{ij}, 0 \leq i \leq m, 0 \leq j \leq m)$ is the variance-covariance matrix of the record values from the standard distribution and $\omega^T = (\sigma_{0,s}, \sigma_{1,s}, \ldots, \sigma_{m,s})$. A detailed explanation of the above equation is provided in Arnold et al. (1998).

Equation (3.4.1.1) has been used to develop predictors for a number of location-scale families. Examples include, but are not limited to, the two-parameter exponential, extreme value, and Pareto distributions (Ahsanullah, 1980, 1993), logistic distribution (Balakrishnan et al., 1995), and the Weibull and Rayleigh distributions (Balakrishnan and Chan, 1994). We consider some of these here.

Example 3.4.1.1. Two-Parameter Exponential (Ahsanullah, 1980)
The density for the two-parameter exponential distribution is given by

$$f(x) = \begin{cases} \dfrac{1}{\sigma}\exp(\dfrac{-(x-\mu)}{\sigma}), & x>\mu,\ \sigma>0 \\ 0, & \text{otherwise.} \end{cases}$$

Ahsanullah (1980) shows that the BLUEs for μ and σ are given by

$$\mu* = \frac{(m+1)x(0) - x(m)}{m}$$

and

$$\sigma* = (x(m) - x(0))/m.$$

Furthermore, Ahsanullah (1980) also establishes that $\alpha(s) = (s+1)$ and $\omega^T \Sigma^{-1}(\mathbf{X} - \mu*\mathbf{1} - \sigma*\boldsymbol{\alpha}) = 0$, so the BLUP of $x(s)$ is given by

$$x*(s) = \frac{(m+1)x(0) - x(m)}{m} + (s+1)\frac{x(m) - x(0)}{m}$$

$$= \frac{1}{m}\big(s\,x(m) - (s-m)x(0)\big).$$

Example 3.4.1.2. Logistic Distribution (Balakrishnan et al. 1995)
The density for the logistic distribution is given by

$$f(x) = \begin{cases} \dfrac{1}{\sigma}\dfrac{\exp(\dfrac{-(x-\mu)}{\sigma})}{\left(1 + \exp(\dfrac{-(x-\mu)}{\sigma})\right)^2}, & -\infty<x<\infty,\ \sigma>0 \\ 0, & \text{otherwise.} \end{cases}$$

Balakrishnan et al. (1995) show that the BLUEs for μ and σ are given by

$$\mu* = \sum_{i=0}^{m} a_i x(i)$$

and

$$\sigma* = \sum_{i=0}^{m} b_i x(i).$$

The coefficients a_i and b_i depend on the means and the variance-co-variance matrix of the record values from the standard logistic distribution. In particular, we have

$$\mathbf{a} = \frac{\alpha' B^{-1}\alpha \mathbf{1}' B^{-1} - \alpha' B^{-1}\mathbf{1}\alpha' B^{-1}}{(\alpha' B^{-1}\alpha)(\mathbf{1}' B^{-1}\mathbf{1}) - (\alpha' B^{-1}\mathbf{1})^2}$$

and

$$\mathbf{b} = \frac{\mathbf{1}' B^{-1}\mathbf{1}\alpha' B^{-1} - \mathbf{1}' B^{-1}\alpha \mathbf{1}' B^{-1}}{(\alpha' B^{-1}\alpha)(\mathbf{1}' B^{-1}\mathbf{1}) - (\alpha' B^{-1}\mathbf{1})^2},$$

where $\alpha' = (\alpha_0, \alpha_1, \ldots, \alpha_m)$ is the mean vector of the record values from the standard logistic distribution, B is the variance-covariance matrix of the record values from the standard logistic distribution, and $\mathbf{1}$ is the vector of ones. As shown in Balakrishnan et al. (1995), the α_is are computed by a recursive formula, $\alpha_0 = 0$ and $\alpha_{k+1} = \alpha_k + \zeta(k+1)$, where $\zeta(.)$ is the Reimann zeta function. The components of B also involve the Reimann zeta integral and are computed by the following formulas.

$$\beta_{k,k} = 2k\zeta(k+1) - k - S_k^2 + 2T_k,$$

where

$$S_k = \sum_{n=1}^{\infty} \frac{1}{n(n+1)^k},$$

$$T_k = \sum_{\ell=2}^{\infty} \frac{1}{\ell(\ell+1)^k}\left(1 + \frac{1}{2} + \cdots + \frac{1}{\ell-1}\right),$$

and

$$\beta_{r,k} = r\{\zeta(r+1) + \zeta(k+1) - 1\} - S_r S_k$$
$$+ \sum_{\ell=1}^{\infty} \frac{1}{\ell(\ell+1)^{k-r}} \sum_{j=1}^{\infty} \frac{1}{j(j+1+\ell)^r} r, \quad r < k.$$

The values of α_k and $\beta_{r,k}$ (for $1 \leq r \leq k \leq 10$) as well as the coefficients a_i and b_i (for values $1 \leq i \leq 10$) have been tabulated by Balakrishnan et al. (1995).

To calculate the BLUP of the next record value, based on the BLUEs once again one uses equation (3.4.1.1). As in Ahsanullah (1980), we also have $\omega^T B^{-1}(X - \mu*1 - \sigma*\alpha) = 0$ for the logistic distribution, so the BLUP of the next record vaules $x(m + 1)$ is simply given by

$$x^*(m+1) = \mu^* + \sigma^* \alpha_m.$$

As an aside, best linear unbiased predictors for *lower records* when the underlying population is the extreme value distribution, normal distribution, or the uniform distribution using generalized least squares theory were also developed independently by Tryfos and Blackmore (1985). Specifically, Tryfos and Blackmore (1985) considered the case where the event of interest was the prediction of athletic records on the basis of past records. Prediction was only done for a time period where one can reasonably assume that the assumptions of the classical record model are satisfied. The bivariate distribution and the covariance of records in any two time periods for any underlying continuous distribution were derived. The results were applied to the special case when the parent distribution was uniform, normal, or uniform to develop the BLUPs of future record values.

3.4.2. Best Linear Invariant Prediction

As noted by Arnold et al. (1998), among others, the best linear invariant predictor of $x(s)$ is based on the result by Mann (1969) and is given by

$$\tilde{x}^*(s) = x^*(s) - \left(\frac{V_4}{1 + V_2}\right)\sigma^*, \qquad (3.4.2.1)$$

where $Var(\sigma^*) = \sigma^2 V_2$ and

$$\sigma^2 V_4 = Cov(\sigma^*, (1 - \omega^T \Sigma^{-1} 1)\mu^* + (\alpha_s - \omega^T \Sigma^{-1}\alpha)\sigma^*).$$

Recall that ω, 1, and Σ were defined in Section 3.4.1.

Ahsanullah (1980) also obtained the BLIP (best linear invariant predictor) of $x(s)$ based on the first $(m + 1)$ record values using equation (3.4.2.1) for the two-parameter exponential.

For the two-parameter exponential, $V_2 = 1/m$ and $V_4 = (s - m)/m$. Substituting these values in (3.4.2.1) along with the expression for $x^*(s)$ gives the BLIP of $x(s)$ as

$$\tilde{x}^*(s) = \frac{(s+1)x(m) - (s-m)x(0)}{m+1}.$$

We note that, similarly, one can also obtain the BLUP and the BLIP for a future *lower* record value based on previous lower record values. As an example, we consider the Type I extreme value distribution here.

Example 3.4.2.1. Type I Extreme Value Distribution (Ahsanullah, 1993) Let $x(0)$, $x(1)$, . . . , $x(m)$ be $(m + 1)$ lower record values from the Type I extreme value distribution with parameters μ and σ (henceforth referred to as EV(μ, σ)). Then the best linear unbiased estimates of μ and σ are given by

$$\mu^* = \frac{\alpha_m}{m} \sum_{i=0}^{m-1} x(i) + (1+\alpha_{m+1})x(m)$$

and

$$\sigma^* = \frac{1}{m} \sum_{i=0}^{m-1} x(i) - x(m),$$

where $\alpha_1 = \nu$, ν is the Euler's constant, and $\alpha_j = \upsilon - \sum_{i=1}^{j-1}(1/i)$.

Using the result by Goldberger (1962) and the fact that $\omega^T \Sigma^{-1}(\mathbf{X} - \mu^*\mathbf{1} - \sigma^*\mathbf{\alpha}) = 0$, we have that the BLUP of $x(s)$ is

$$x^*(s) = \mu^* + \sigma^* \alpha_{s+1}.$$

Similarly equation (3.4.2.1) gives the BLIP of $x(s)$ as

$$\tilde{x}^*(s) = x^*(s) - \frac{\alpha_{s+1} - \alpha_{m+1}}{m+1}\sigma^* = \mu^* + \alpha_{s+1}\sigma^* - \frac{\alpha_{s+1} - \alpha_{m+1}}{m+1}\sigma^*.$$

3.4.3. "Other" Optimal Predictors

In addition to studying the development of BLUPs and BLIPs, Ahsanullah (1980, 1993) also considered best least square predictors of future record values. The best (unrestricted) least squares predictor of the future record $x(s)$, $\hat{x}^*(s)$, based on the first $(m + 1)$ records is given by

$$\hat{x}^*(s) = E(x(s)| x(0), x(1), . . ., x(m)).$$

Ahsanullah calculated $\hat{x}*(s)$ for records from the two-parameter exponential and the generalized Pareto distribution.

For the two-parameter exponential, the best unrestricted least squares predictor is

$$\hat{x}*(s) = x(m) + (s - m)\sigma. \tag{3.4.3.1}$$

Note that the above expression depends on the unknown parameter σ. If we substitute the BLUE of σ in (3.4.3.1), then $\hat{x}*(s)$ becomes equal to the BLUP of $x(s)$. Similarly, when the best unrestricted predictor is calculated for the generalized Pareto distribution, its value also depends on unknown parameters.

Asymptotic linear prediction of future record values is another area of interest as identified by Arnold et al. (1998). Specifically, they consider the work of Nagaraja (1984) who studied the prediction of the sth order statistic using r lower order statistics from a large sample, under the assumption that the appropriately normalized sample minimum $(y_{(1)})$ converges to a degenerate distribution G as the sample size approaches infinity. In other words, assume that for a sample of size n, there exist constants c_n and $d_n > 0$, such that

$$P\left(\frac{Y_{(1)} - c_n}{d_n} \leq y\right) \to G(y)$$

as $n \to \infty$ for all y in the support of G. Then Nagaraja (1984) shows that the r order statistics behave as the first r upper record values from the distribution $G(\mu + \sigma y)$, where $\mu = c_n$ and $\sigma = d_n$. So in fact, under the above situation, the problem now reduces to that of predicting the sth record value given r upper record values. As an aside, from Gnedenko (1943) and as noted in Arnold et al. (1998), G can be one of the Frechét, Weibull, or extreme-value distributions. Moreover, the constants c_n and d_n depend on F and G and are available in a number of books, for example, Galambos (1978, pp. 56–57).

3.4.4. Prediction Intervals

A discussion on prediction of future record values would be incomplete without a subsection on prediction intervals. Work in this area was started by Dunsmore (1983) with the development of Bayesian prediction intervals and tolerance region prediction for the difference in two future record values for the one-parameter and the two-parameter exponential distribution. Some of this work is detailed in the next chapter on Bayesian inference from record-breaking data. Work on prediction intervals has, since then, been continued by Balakrishnan et al. (1995), Chan (1998), and Berred (1998) among others. Since the work of Chan (1998) deals with the prediction of future record values for any location-scale family, we start by presenting it and showing how it applies to work in prediction of future record values. The work by Berred (1998) has not been reviewed in the book by Arnold et al. (1998) and therefore serves as the conclusion of this subsection.

Example 3.4.4.1. Location-Scale Family (Chan, 1995)
Assume as in Section 3.4.1, that the underlying cumulative distribution function F belongs to a location-scale family, with location parameter μ and scale parameter σ. That is, $F(y) = G((y - \mu)/\sigma)$, where $G(\cdot)$ is the c.d.f. for the standard family ($\mu = 0$, $\sigma = 1$). Let $(\tilde{\mu}, \tilde{\sigma})$ be any pair of equivariant estimators for (μ, σ) (e.g., MLEs or BLUEs) based on the first $(m + 1)$ record values. From Chan (1995), we have that the quantity $Z_3 = (x(m+1) - x(m))/\tilde{\sigma}$ serves as a pivotal in constructing a confidence interval for the next record value, $x(m + 1)$. This is based on the following theorem by Chan (1998).

Theorem 3.4.4.1: Let $(\tilde{\mu}, \tilde{\sigma})$ be any pair of equivariant estimators for (μ, σ) based on the first $(m + 1)$ record values drawn from a cumulative distribution function F from a location-scale family. Let $Z_1 = (\tilde{\mu} - \mu)/\tilde{\sigma}$, $Z_2 = \tilde{\sigma}/\sigma$, and $Z_3 = (x(m+1) - x(m))/\tilde{\sigma}$. Also, define $a_i = (x(i) - \tilde{\mu})/\tilde{\sigma}$, $i = 0, 1, \ldots, m$. Then the conditional prediction density of Z_3 ($0 < z_3 < \infty$) is of the form

$$g(z_3 \mid \mathbf{a}) =$$

$$\left(C(\mathbf{a}, m+1) \int_{-\infty}^{\infty} \int_{0}^{\infty} z_2^{m+1} g((z_3 + a_m)z_2 + z_1 z_2) \prod_{i=0}^{m-1} \frac{g(a_i z_2 + z_1 z_2)}{1 - G(a_i z_2 + z_1 z_2)} \, dz_2 dz_1 \right),$$

where C is a function of the ancillary statistics, $a_1, a_2, \ldots, a_m$.

Chan (1998) used Theorem 3.4.4.1 to develop prediction intervals for a future record value from the Gumbel distribution. This theorem was also used by Balakrishnan et al. (1995) and by Balakrishnan and Chan (1998) to develop prediction intervals for future record values from the logistic and the normal distributions, respectively.

Example 3.4.4.2.
The work of Berred (1998) proposes a method to find tolerance regions for future record values when the underlying distribution has a

regularly varying tail. In other words, the survival function $\overline{F}(x)$ is given by

$$\overline{F}(x) = x^{-1/\alpha} L(x), \qquad x \geq \alpha, \tag{3.4.4.1}$$

where $\alpha \geq 0$ and L is a slowly varying function at infinity. As before, assume that m record values have been observed. Berred (1998) proposes using the following statistic to develop a tolerance region for the $(m + r)$th record value

$$\gamma_{r,m} = \frac{\log x(r+m) - \log x(m)}{\hat{\alpha}}. \tag{3.4.4.2}$$

Here

$$\hat{\alpha} = \frac{1}{km - (k(k-1)/2)} \sum_{i=1}^{k} \log x(m-i+1)$$

(for some $1 \leq k < m$) is an estimator of α as proposed in Berred (1992). Berred shows that the statistic defined in (3.4.4.2) has an asymptotic gamma distribution (as the number of records m goes to infinity) with parameters r and 1. Hence, tolerance regions can be easily constructed.

As an example, a tolerance region with $1 - \beta$ coverage probability for $\log(x(r + m)/x(m))$ will have the form $[0, q_{1-\beta}^r \hat{\alpha}]$, where $q_{1-\beta}^r$ is the $(1 - \beta)$th percentile of the gamma distribution with parameters r and 1.

Furthermore, under certain additional conditions on the tail of F and when both r and m go to infinity, Berred (1998) shows that the statistic

$$T_{r,m} = m^{-1/2} \left(\frac{\log x(r+m) - \log x(m)}{\hat{\alpha}} - m \right)$$

converges in distribution to the standard normal distribution. Thus, one can use the percentiles of the standard normal distribution to construct tolerance prediction regions under these conditions.

Some examples of the underlying distributions that satisfy (3.4.4.1) are:

a) $\overline{F}(x) = x^{-1/\alpha} (b + dx^{-\theta})$ for large x and $b, d, \theta > 0$

b) $\overline{F}(x) = b x^{-1/\alpha} (\log x)^\theta$ for large x, $b > 0$, and $\theta \neq 0$.

4
Nonparametric Inference—Genesis

4.1 Introduction

In doing statistical inference, if you do not have to make assumptions about the underlying population being sampled, then it is best to not do so. Record-breaking data often arise in industrial settings, and therefore it seems natural to use the exponential, extreme value, or Weibull distributions as models for the underlying populations of interest. However, in general, there is no guarantee that the correct parametric model is being used in order to make inferences from the data. So, there was a real need to develop nonparametric inference from such data.

Nonparametric inference from record values saw its birth in the paper by Foster and Stuart (1954). They proposed two very simple distribution-free statistics to test for randomness in a series of observations using upper and lower records. However, after that, work on all inference from records apparently came to a standstill, only to resume some twenty years later with work on prediction of future records. General nonparametric inference did not appear until 1988 when Samaniego and Whitaker published a paper on nonparametric maximum likelihood estimation from such data. They developed and studied the nonparametric MLE of the underlying distribution function and laid the foundation for further nonparametric work in this field. After that, there were several papers on smooth function estimation, and even on Bayesian nonparametric inference, from record-breaking data. Some of these results are presented in the following sections and chapters.

4.2 The Work of Foster and Stuart

The statistics proposed by Foster and Stuart (1954) are linear functions of the upper records (or of the lower records) obtained from a series of observations. To test for a trend in location, the proposed statistic is the difference of the number of upper records and the number of lower records in a series. Similarly, to test for trend in dispersion, the proposed statistic is the sum of the number of upper and lower records in the series. It is intuitively clear that if there is a trend in location, there will tend to be more records in one direction than the other, and if there is an increasing (decreasing) trend in dispersion, there will be too many (or too few) records. Under the null hypothesis, Foster and Stuart provided a joint distribution of the statistics and showed that the statistics are uncorrelated and asymptotically normally distributed. We discuss their test statistics next.

Suppose as before that $Y_1, Y_2, \ldots, Y_n$ is a random sample from a continuous distribution function F with density f. Define the following scores.

$$u_r = \begin{cases} 1, & \text{if the } r\text{th observation is an upper record} \\ 0, & \text{otherwise,} \end{cases}$$

and

$$l_r = \begin{cases} 1, & \text{if the } r\text{th observation is a lower record} \\ 0, & \text{otherwise.} \end{cases}$$

Furthermore, let

$$s_r = u_r + l_r$$

and

$$d_r = u_r - l_r.$$

Then the test statistics are

$$s = \sum_{r=2}^{n} s_r, \quad \text{with } 0 \leq s \leq n - 1,$$

and

$$d = \sum_{r=2}^{n} d_r, \quad \text{with } -(n - 1) \leq d \leq n - 1. \tag{4.2.1}$$

Under the null hypothesis that the data are i.i.d., every ordering of a series of r observations is equally probable. Hence,

$$P(u_r = 1) = P(l_r = 1) = 1/r \text{ and } P(u_r = 0, l_r = 0) = 1 - \frac{2}{r}. \quad (4.2.2)$$

We denote by $p^{(r)}(s,d)$, the joint distribution of s and d in a series of r observations. Since the probability that the rth observation is a record is independent of the order (among themselves) of the preceding observations, it follows that

$$p^{(r)}(s,d) = \left(1 - \frac{2}{r}\right) p^{(r-1)}(s,d) + \frac{1}{r} p^{(r-1)}(s-1, d-1) + \frac{1}{r} p^{(r-1)}(s-1, d+1),$$

$$(4.2.3)$$

subject to the initial condition $p^{(1)}(0,0) = 1$.

Using the Fourier inversion theorem and (4.2.3), the marginal density of s is obtained as

$$p_1^{(n)}(s) = \frac{1}{n!} 2^s u^{(n-2)} (n-s-1), \quad (4.2.4)$$

where $u^{(n)}(r)$ is the sum of the products of all selections of r integers out of $1, 2, \ldots, n$. The joint density of s and d is then

$$p^{(n)}(s,d) = p_1^{(n)}(s) 2^{-s} \binom{s}{\frac{1}{2}(s+d)}, \quad s = 0, \ldots, n-1; \quad (4.2.5)$$

$$d = -(n-1), \ldots, (n-1).$$

The marginal frequency of d can be obtained by summing over s in (4.2.5). Foster and Stuart also show that there is no correlation between s and d and further derive the asymptotic distribution of the statistics. It is shown that s and d are asymptotically normally distributed and independent. The moments of the statistics are given by

$$E(s) = \mu = 2 \sum_2^n \frac{1}{r},$$

$$Var(s) = 2 \sum_2^n \frac{1}{r} - 4 \sum_2^n \frac{1}{r^2},$$

$$E(d) = 0, \text{ and } Var(d) = \mu.$$

They also generate the exact distribution of s and d for small sample sizes. Tables of the exact percentiles can be found in either the article of Foster and Stuart or in the book by Arnold et al. (1998). Foster and Stuart show that the tests are consistent against the alternate hypothesis of trend in the data set.

4.3 Nonparametric Maximum Likelihood Estimation

Assuming the case of random sampling, the original sample consists of n observations, $Y_1, Y_2, \ldots, Y_n$, and the available record-breaking data are $R, X_1, K_1, X_2, K_2, \ldots, X_R, K_R$. The likelihood function of the data was derived in Chapter 3 and is given by

$$L = \prod_{i=1}^{r} f(x_i)(1 - F(x_i))^{k_i - 1} . \tag{4.3.1}$$

The nonparametric maximum likelihood estimator of F is then the function F^*, which maximizes (4.3.1) rewritten as

$$L = \prod_{i=1}^{r} (F(x_i) - F(x_i^-))(1 - F(x_i))^{k_i - 1} . \tag{4.3.2}$$

Any function that maximizes (4.3.2) places all of its mass on the interval (x_1, ∞) together with the record values $x_1, x_2, \ldots, x_r$, themselves. Samaniego and Whitaker (1988) suggest reparameterizing the likelihood function (4.3.2) in terms of the survival probabilities,

$$\Phi_i = \frac{1 - F(x_i)}{1 - F(x_{i-1})} ,$$

where $x_0 = 0$, and $x_1 < x_2 < \cdots < x_r$ are the record values. It is clear then, that (4.3.2) can be rewritten as

$$L = L(\Phi) = \prod_{i=1}^{r} (1 - \Phi_i)\Phi_i^{\sum_{j=i}^{r} k_j - 1} , \tag{4.3.3}$$

where the sequence $\{k_i\}$, $i = 1, 2, \ldots, r$, is the interrecord times for the sequence $\{x_i, i = 1, 2, \ldots, r\}$.

Using standard methods to find maximum likelihood estimates then gives us that $L(\Phi)$ is maximized by the vector $\Phi = (\Phi_1, \Phi_2, \ldots, \Phi_r)$ defined by

$$\Phi_i = \frac{\sum_{j=i}^{r} k_j - 1}{\sum_{j=i}^{r} k_j}. \tag{4.3.4}$$

Substituting in (4.3.2) gives the nonparametric MLE of the survival function as

$$\bar{F}^*(t) = \prod_{\{i:\, X_i \le t\}} \frac{\sum_{j=i}^{r} K_j - 1}{\sum_{j=i}^{r} K_j}. \tag{4.3.5}$$

Samaniego and Whitaker note that the above estimator bears some resemblance to the Kaplan–Meier estimator of the survival function under randomly right-censored data (Kaplan and Meier, 1958). To see the analogy, define the ith interval of observation as $I_i = [x_i , x_{i+1})$. There are $N_i = \sum_{j=i}^{r} k_j$ individuals at risk and $D_i = 1$ failure in this interval, with $\sum_{j=i}^{r} k_j - 1$ individuals withdrawn from the study at the end of this interval. With this identification, the estimator in (4.3.5), $\bar{F}^*(t) = \prod_{\{i:X_i \le t\}} (N_i - D_i)/N_i$, is exactly the form of the Kaplan –Meier estimator. The one difference between the Kaplan–Meier estimator and the one in (4.3.5) is that the estimator in (4.3.5) does not decrease to zero as t increases to infinity, and the Kaplan–Meier estimator is zero beyond the last observation if that observation is uncensored (or if the "self-consistent" estimator is used).

Samaniego and Whitaker (1988) also performed a limited simulation study that focused on the problem of estimating small population quantiles. They examined the behavior of the nonparametric estimator of small population quantiles and compared it to the behavior of its parametric counterpart, assuming an underlying exponential distribution. In other words, if $t_p = F^{-1}(p)$ is the required population quantile, they compared the mean-squared errors of $\hat{T}_1 = F^{*-1}(p)$ and $\hat{T}_2 = -\tilde{\mu} \ln(1-p)$, where $\tilde{\mu} = \sum_{i=1}^{R} K_i X_i / R$ is the estimator of the mean of an underlying exponential distribution as defined in Chapter 3. They considered samples of size $n = 20$, 50, and 100 from three different distributions, the exponential with mean one and the Weibull distributions with shape parameters 2 and 4 (the scale was always one). They found that when one sampled from an exponential distribution, the relative efficiency of $\hat{T}_2$ with respect to $\hat{T}_1$ was between 20 and 80%. Under mild departures from exponentiality (for example, the Weibull distribution with shape parameter 2), the two estimators were somewhat similar in their performance, with the nonparametric estimator having a certain edge. The relative performance of $\hat{T}_2$, however, deteriorated rapidly with increasing departure from exponentiality. Moreover, in extreme departures from exponentiality, $\hat{T}_2$ also seriously underestimated the required quantiles. Table 4.3.1 gives results from their simulations for the Weibull distribution with shape parameter 4.

The problem with a single sample is that, since the data values represent only weaker or smaller values in general, they can provide reliable information only in the left tail of F. In fact, Gulati (1991) has shown that F^* is consistent only in the extreme left tail of the underlying distribution. Thus, F^* can estimate F efficiently only in the parametric context. A nonparametric approach seems promising only in the following situations.

   a) F^* is used to estimate small population quantiles, or
   b)   the process of observing records can be repeated.

Table 4.3.1. Efficiency of $\hat{T}_2$ vs. $\hat{T}_1$ for the Weibull (1, 4)

n	$1 - F(T)$	T	$AV(\hat{T}_1)$	$MSE(\hat{T}_1)$	$AV(\hat{T}_2)$	$MSE(\hat{T}_2)$	$RE(\hat{T}_2/\hat{T}_1)$
20	0.90	0.5697	0.5028	0.0222	0.3716	0.0724	3.2590
50	0.90	0.5697	0.4810	0.0234	0.6041	0.0932	3.9840
100	0.90	0.5697	0.4695	0.0229	0.8685	0.2371	10.3520
20	0.95	0.4759	0.4336	0.0171	0.1809	0.0949	5.5400
50	0.98	0.3770	0.3425	0.0107	0.1158	0.0716	6.6670
100	0.99	0.3166	0.2899	0.0066	0.0828	0.0560	8.5360

Situation (a) is useful in the estimation of strength of materials for engineering design and is a special case of the second situation. Therefore, to estimate F nonparametrically, Samaniego and Whitaker (1988) extended the single sample results to the multisample case. We now assume that m independent samples of size n are obtained sequentially from F and we only record successive minimum values in each. Hence our observed data are

$$X_{i1}, K_{i1}, \ldots, X_{iR_{(i)}}, K_{iR_{(i)}}, \quad i = 1, \ldots, m,$$

where $R_{(i)}$ denotes the number of record values from the ith sample.

The derivation of the maximum likelihood estimator for a single sample generalizes easily to the case of m independent sequences. The likelihood is now written as

$$L = \prod_{i=1}^{m} \prod_{j=1}^{r_i} (F(x_{ij}) - F(x_{ij}^{-}))(1 - F(x_{ij}))^{k_i - 1} \quad . \tag{4.3.6}$$

Let $\{x_{(i)}, i = 1, 2, \ldots, r^*\}$ denote the ordered values of the m samples combined so that $r^* = \Sigma r_i$ and let $\{k_{(i)}, i = 1, 2, \ldots, r^*\}$ denote the k_{ij}s corresponding to the ordered $x_{(i)}$s. Now, following the same reparameterization as for the single sample case, the likelihood can be rewritten as

$$L = L(\Phi) = \prod_{i=1}^{r^*} (1 - \Phi_i) \Phi_i^{\sum_{j=i}^{r^*} k_{(j)} - 1} \quad .$$

Using the same methods as before, it then follows that the nonparametric MLE of the survival function is now given by

$$\bar{F}_{mn}(t) = \prod_{\{i: X_{(i)} \leq t\}} \frac{\sum_{j=i}^{r^*} K_{(j)} - 1}{\sum_{j=i}^{r^*} K_{(j)}} \quad . \tag{4.3.7}$$

Next, we examine the asymptotic behavior of the estimator (4.3.7) as $m \to \infty$. The details are given by Samaniego and Whitaker (1988).

4.4 Asymptotic Results

Since F_{mn} has a multiplicative form, it is more convenient to study the asymptotic behavior of

$$\Lambda_{mn}(t) = -\ln(1 - F_{mn}(t)) = -\ln(\overline{F}_{mn}(t))$$

as an estimator of the theoretical hazard function $\Lambda(t) = -\ln(1 - F(t))$. To do so, we establish a representation of the hazard function in terms of the random records framework. For the record-breaking data as described at the beginning of Section 4.3, define the following quantities: $\kappa_i(x)$ is the number of Y_{ij} known to exceed x on the basis of observed records, and $\mathcal{N}_i(x)$ is the number of record-breaking observations that are less than or equal to x; that is,

$$\kappa_i(x) = \sum_{j=1}^{n} I\left[\min(Y_{i1}, Y_{i2}, \dots, Y_{ij}) > x\right] \tag{4.4.1}$$

and

$$\mathcal{N}_i(x) = \sum_{j=1}^{n} I\left[Y_{ij} \leq x, \, Y_{ij} = \min(Y_{i1}, Y_{i2}, \dots, Y_{ij})\right]. \tag{4.4.2}$$

By virtue of the independence of the replications, it is clear that $\kappa_1(x), \kappa_2(x), \dots, \kappa_m(x)$ are i.i.d. for a fixed sample size n, as are $\mathcal{N}_1(x), \mathcal{N}_2(x), \dots, \mathcal{N}_m(x)$. We then have the following lemma.

Lemma 4.4.1: Let Λ be the theoretical hazard function corresponding to the distribution function F on $(0, \infty)$. Then for all t such that $F(t) < 1$ and for arbitrary $i \in \{1, 2, \dots, m\}$, we have

$$\Lambda(t) = \int_0^t \frac{1}{E(\kappa_i(x))} \, dE(\mathcal{N}_i(x)),$$

where $\kappa_i(x)$ and $\mathcal{N}_i(x)$ were defined in (4.4.1) and (4.4.2).

Proof: From (4.4.1) and (4.4.2), we evaluate the expected values of $\kappa_i(x)$ and $\mathcal{N}_i(x)$ as follows.

$$E\,\kappa_i(x) = \sum_{j=1}^{n} P\left[\min(Y_{i1}, Y_{i2}, \dots, Y_{ij}) > x\right] = \sum_{j=1}^{n} \overline{F}^j(x). \tag{4.4.3}$$

Similarly, it follows that

$$E\,\mathcal{N}_i(x) =$$

$$\sum_{j=1}^{n} P\left[Y_{ij} \leq x \middle| Y_{ij} = \min(Y_{i1}, Y_{i2}, \ldots, Y_{ij})\right] \times P(Y_{ij} = \min(Y_{i1}, Y_{i2}, \ldots, Y_{ij}))$$

$$= \sum_{j=1}^{n} \frac{1 - \overline{F}^j(x)}{j}. \tag{4.4.4}$$

From (4.4.4) we then obtain

$$\frac{dE\mathcal{N}_i(x)}{dF(x)} = \sum_{j=1}^{n} \overline{F}^{\,j-1}(x).$$

Hence, we have the result

$$\int_0^t \frac{1}{E(\kappa_i(x))} dE(\mathcal{N}_i(x)) = \int_0^t \frac{1}{\overline{F}(t)} dF(x) = -\ln(1 - F(t)) = \Lambda(t). \quad \blacklozenge$$

From Lemma 4.4.1, we immediately obtain an expression for a simple estimator for $\underline{\Lambda}(t)$. Consider the respective averages of the sequences $\{\kappa_i(x), i = 1, 2, \ldots, m\}$ and $\{\mathcal{N}_i(x), i = 1, 2, \ldots, m\}$:

$$\overline{\kappa}(x) = \frac{1}{m} \sum_{i=1}^{m} \kappa_i(x) \quad \text{and} \quad \overline{\mathcal{N}}(x) = \sum_{i=1}^{m} \mathcal{N}_i(x).$$

A natural estimator of $\Lambda(t)$ is therefore given by

$$\Lambda^*(t) = \int_0^t \frac{1}{\overline{\kappa}(x)} d\overline{\mathcal{N}}(x). \tag{4.4.5}$$

Note that $\overline{\kappa}(x)$ and $\overline{\mathcal{N}}(x)$ are strong consistent estimators of $E\mathcal{N}(x)$ and $E[\kappa(x)]$. Using the Gilvenko–Cantelli theorem, one can then establish that Λ^* is a strong uniform consistent estimator of Λ. Next, we show that Λ_{mn} and Λ^* are asymptotically equivalent.

Lemma 4.4.2: Let $\{Y_{ij}, 1 \leq j \leq n, 1 \leq i \leq m\}$ be m independent samples of size n from the distribution F, and let $\{(X_{ij}, K_{ij}), 1 \leq j \leq R_i, 1 \leq i \leq m\}$ be the random record data drawn from the $\{Y_{ij}\}$. Let T be such that $F(T) < 1$. Then

$$\sup_{0 \leq t \leq T} \sqrt{m} \, |\Lambda_{mn}(t) - \Lambda^*(t)| \to 0 \quad \text{with probability one as } m \to \infty.$$

Proof: To avoid trivialities, we assume that $F(T) > 0$. Let $t \in (0, T)$. We first rewrite Λ_{mn} in a form comparable to Λ^* as in (4.4.5). From (4.3.7), we have

$$\Lambda_{mn}(t) = - \sum_{\{i:\, X_{(i)} \leq t\}} \ln \frac{\displaystyle\sum_{j=i}^{r^*} K_{(j)} - 1}{\displaystyle\sum_{j=i}^{r^*} K_{(j)}} .$$

From the definition of $\{\kappa_i(.)\}$ we have that

$$\sum_{j=1}^{r^*} K_{(j)} = \sum_{l=1}^{m} \kappa_l(x_{(i)}) + 1.$$

Thus,

$$\Lambda_{mn}(t) = \sum_{\{i:\, x_{(i)} \leq t\}} \ln \frac{\displaystyle\sum_{l=1}^{m} \kappa_l(x_{(i)}) + 1}{\displaystyle\sum_{l=1}^{m} \kappa_l(x_{(i)})}$$

$$= \int_0^t m \ln \left[\frac{\displaystyle\sum_{l=1}^{m} \kappa_l(x_{(i)}) + 1}{\displaystyle\sum_{l=1}^{m} \kappa_l(x_{(i)})} \right] d\,\overline{\mathcal{N}}(x).$$

We now restrict ourselves to the event $\{X_{(R^*)} > T\}$, the occurrence of which ensures that both Λ^* and Λ_{mn} are well defined in $[0, T]$. Since, with probability one, the event $\{X_{(R^*)} > T\}$ occurs for m sufficiently large, the restriction made above does not limit the generality of the result. Using the inequality

$$\ln\left(\frac{a+1}{a} \right) < \frac{1}{a} + \frac{1}{a(a+1)},$$

we have

$$|\Lambda_{mn}(t) - \Lambda^*(t)| = \left| \int_0^t \left\{ m \ln\left[\frac{\sum \kappa_l(x) + 1}{\sum \kappa_l(x)} \right] - [\kappa(x)]^{-1} \right\} d\overline{\mathcal{N}}(x) \right|$$

$$< \int_0^t m \left[\frac{1}{\left(\sum \kappa_l(x)\right)\left(\sum \kappa_l(x) + 1\right)} \right] d\, \overline{\mathcal{N}}(x)$$

$$< \int_0^t m \left[\frac{1}{\left(\sum \kappa_l(x)\right)} \right]^2 d\, \overline{\mathcal{N}}(x)$$

$$< m \left[\frac{1}{\left(\sum \kappa_l(x)\right)} \right]^2 \int_0^t d\, \overline{\mathcal{N}}(x)$$

$$= \frac{1}{m\, \overline{\kappa}^2(x)} \overline{\mathcal{N}}(t) . \qquad (4.4.6)$$

Since the function in (4.4.6) is, with probability one, a non-decreasing function of t, we have that

$$\sup_{0 \leq t \leq T} |\Lambda_{mn}(t) - \Lambda^*(t)| < \frac{1}{m\, \overline{\kappa}^2(x)} \overline{\mathcal{N}}(t) \quad \text{with probability one.}$$

Hence, the strong law of large numbers gives

$$\sup_{0 \leq t \leq T} |\Lambda_{mn}(t) - \Lambda^*(t)| \to 0 \quad \text{with probability one as } m \to \infty. \quad \blacklozenge$$

Since Λ_{mn} and Λ^* are asymptotically equivalent and Λ^* is a strong consistent estimator of Λ, the true hazard function, then Λ_{mn} must be a strong uniform consistent estimator of Λ. The strong consistency of F_{mn} then follows from the strong consistency of Λ_{mn}. The next theorem is stated without proof. For details, the reader is referred to Samaniego and Whitaker (1988).

Theorem 4.4.1: (Strong Uniform Consistency of F_{mn}) As in Lemma 4.4.2, let $\{Y_{ij}, 1 \leq j \leq n, 1 \leq i \leq m\}$ be m independent samples of size n from the distribution F, and let $\{(X_{ij}, K_{ij}), 1 \leq j \leq R_i, 1 \leq i \leq m\}$ be the random record data drawn from the $\{Y_{ij}\}$. Further let F_{mn} be the nonparametric MLE of the distribution function. Then

$$\sup_{t \geq 0} |F_{mn}(t) - F(t)| \to 0 \quad \text{with probability one as } m \to \infty .$$

The asymptotic theory for F_{mn} can be derived using the methods of Andersen and Gill (1982). The theorem is once again stated without proof.

Theorem 4.4.2: (Asymptotic Normality of F_{mn}) For all T such that $F(T) < 1$, the process $\left\{ m^{1/2}(\bar{F}_{mn}(t) - \bar{F}(t)) : 0 \leq t \leq T \right\}$ converges weakly to a mean zero Gaussian process with covariance function

$$R(s,t) = \bar{F}(s)\bar{F}(t) \int_0^t \frac{1}{E(\kappa(t))} \, d\Lambda(x),$$

where $\kappa(x)$ has been defined earlier.

Samaniego and Whitaker (1988) also mention that even though the notation and arguments are more cumbersome, it is possible to derive the nonparametric MLE and to argue similar asymptotic behavior when the sample size is allowed to vary. It is also possible to obtain similar results under inverse sampling, as demonstrated later.

5
Smooth Function Estimation

Nonparametric estimation for lifetime or continuous data has its roots in the fifteenth century. However, until the 1950s, all nonparametric estimators obtained for the probability density function and the cumulative distribution function were discrete. Typically, these estimates were hard to update and required more computation than necessary. This was not a serious problem if the functions being estimated were themselves discrete, but most lifetime or measurement data are continuous and so "smoother" estimators were needed.

Rosenblatt (1956) used the histogram estimator of the density function to develop the first "smooth" density estimator. Rosenblatt's estimator was simply a "shifted" histogram. In other words, to estimate the density function at a point x, one shifts the classical histogram so that $\underline{x}$ lies in the center of a "mesh" interval. To calculate the density at another point, say, y, the mesh is shifted again so that now y is in the center of a mesh interval. Later, a detailed account and some properties of "smooth" density estimators were provided by Parzen (1962). For a positive "bandwidth" sequence h_n and a suitable "kernel function" K, Parzen considered as an estimate of the density function $f(x)$ the following function,

$$f_n(x) = \int_{-\infty}^{\infty} \frac{1}{h_n} K((x-t)/h_n) dF_n(t). \tag{5.1}$$

Here F_n is the empirical cumulative distribution function based on a random sample of size n from the density f. When the data consist of n i.i.d. observations, $x_1, x_2, x_3, \ldots, x_n$, the kernel density function estimator in (5.1) can be written as a relatively simple sum

$$f_n(x) = \frac{1}{nh_n} \sum_{j=1}^{n} K((x-x_j)/h_n). \tag{5.2}$$

Parzen (1962) showed that the estimator in (5.2) is consistent and under certain regularity conditions is asymptotically normal as well. It has also been shown that for a suitable kernel function and bandwidth sequence, the kernel density estimator $f_n(x)$ has a smaller mean-squared error than its empirical counterpart, the histogram estimator. Since the development of kernel density estimators by Parzen, kernel-based smooth estimation has been studied for various types of data and functions other than the density function. In particular, distribution functions, quantile functions, and hazard functions have been investigated in the literature, and i.i.d. samples have been extended to various types of incomplete or truncated data.

As mentioned earlier, record-breaking data are typically contin-uous in nature. The estimator F_{mn} of the distribution function given by (4.3.7) is a step function and once again a smooth estimator of the c.d.f., F, is of interest. Gulati and Padgett (1992, 1994(a,b,c)) applied the ideas of Parzen to record-breaking data and used F_{mn} to develop "smooth" estimators for the underlying distribution, density, quantile, hazard, and hazard-rate functions. We present these estimators in this chapter along with their properties.

5.1 The Smooth Function Estimators—Definitions and Notation

As in Chapter 4, the observed data consist of m independent identically distributed sequences of record-breaking data:

$$X_{i1}, K_{i1}, \ldots, X_{iR_{(i)}}, K_{iR_{(i)}}, \quad i = 1, \ldots, m.$$

In looking at smooth estimation from record-breaking data, first consider the kernel-type estimation of the underlying distribution function F. For a suitable kernel function K and a positive bandwidth sequence h_{mn}, this estimator, denoted by $\hat{F}_{mn}$, is defined as the kernel-smoothed version of F_{mn}, given by

$$\hat{F}_{mn}(x) = \int_0^\infty \frac{1}{h_{mn}} K((x-t)/h_{mn}) F_{mn}(t)dt$$

$$= \sum_{j=1}^{r^*} F_{mn}(X_{(j)}) \int_{X_{(j)}}^{X_{(j+1)}} \frac{1}{h_{mn}} K((x-t)/h_{mn}) \, dt , \qquad (5.1.1)$$

where, as before, $\{X_{(i)}, i = 1, 2, \ldots, r^*\}$ denotes the ordered values of the m samples combined, so that $r^* = \Sigma r_i$ and $\{K_{(i)}, i = 1, 2, \ldots, r\}$ denotes the K_{ij}s corresponding to the $X_{(i)}$s. Also, we define $X_{(r^*+1)}$ to be ∞.

One can now differentiate $\hat{F}_{mn}(x)$ to obtain a smooth estimator for the density $f(x)$. Denoted by $f_{mn}(x)$, this estimator is given by

$$f_{mn}(x) = \frac{d}{dx}\hat{F}_{mn}(x)$$

$$= \int_0^\infty \frac{1}{h_{mn}^2} K'((x-t)/h_{mn}) F_{mn}(t)dt,$$

where the derivative is defined in the usual sense and K' denotes the derivative of K. By expanding the above integral, one can easily see that $f_{mn}(x)$ is analogous to the usual kernel density estimator for the i.i.d. sample; that is,

$$f_{mn}(x) = \int_0^\infty \frac{1}{h_{mn}^2} K'((x-t)/h_{mn}) F_{mn}(t)dt$$

$$= \sum_{j=1}^r F_{mn}(X_{(j)})h_{mn}^{-1}\left[K((x-X_{(j)})/h_{mn}) - K((x-X_{(j+1)})/h_{mn})\right]$$

$$= \sum_{j=1}^r S_j h_{mn}^{-1} K((x-X_{(j)})/h_{mn}), \tag{5.1.2}$$

where S_j denotes the jump of F_{mn} at $X_{(j)}$.

Gulati and Padgett (1994b) also consider three estimators of the quantile function $Q(p) = F^{-1}(p)$, $0 < p < 1$. The simplest of these is the *"empirical quantile function."* Denoted by $Q_{mn}(p)$, it is a step function defined by

$$Q_{mn}(p) = F_{mn}^{-1}(p) = \inf\{y \mid F_{mn}(y) \geq p\}. \tag{5.1.3}$$

Next, we define two "smooth" estimators for the quantile function. Given $Q_{mn}(p)$, a kernel-type estimator of $Q(p)$ for a suitable kernel function $K(\cdot)$, and a positive bandwidth sequence h_{mn}, is defined as

$$\hat{Q}_{mn}(x) = \int_0^1 \frac{1}{h_{mn}} K((x-t)/h_{mn}) Q_{mn}(t)dt$$

$$= \sum_{j=1}^r X_{(j)} \int_{t_{(j-1)}}^{t_{(j)}} \frac{1}{h_{mn}} K((x-t)/h_{mn}) \, dt, \tag{5.1.4}$$

where $t_{(j)} = F_{mn}(X_{(j)})$.

The third and final estimator of the quantile function $x_{mn}(p)$ is the *Nadaraya-type estimator* and is obtained by inverting the smooth c.d.f. estimator $\hat{F}_{mn}(x)$; that is, $x_{mn}(p)$ for a given p, is the solution for x in the equation $\hat{F}_{mn}(x) = p$.

Finally, we consider hazard function estimation. The estimators of the (cumulative) hazard and the hazard rate functions are defined in a manner similar to the quantile function estimators. Let $T_F = \sup\{t \mid F(t) < 1\}$ and let $0 < T^* < T_F$ be arbitrarily close to T_F such that $T^* < \infty$. Then for $0 \le x \le T^*$, the following three estimators of the hazard function, $\Lambda(x) = -\ln(1 - F(x))$, are defined.

a) *Empirical Hazard Function Estimator*: Denoted by $\Lambda_{mn}(x)$, as discussed in Chapter 4, this estimator was defined by Samaniego and Whitaker (1988) as

$$\Lambda_{mn}(x) = -\ln(1 - F_{mn}(x)) . \tag{5.1.5}$$

Recall that they established strong uniform consistency for F_{mn} and, hence, for this estimator of the hazard function.

b) *Kernel-Type Estimator*: For a suitable kernel function $K(\cdot)$, and a positive bandwidth sequence h_{mn}, the kernel-type smooth hazard function estimator is defined as

$$\Lambda_{mn}^*(x) = \int_0^{T^*} \frac{1}{h_{mn}} K((x-t)/h_{mn}) \, \Lambda_{mn}(t) dt$$

$$= \sum_{j=1}^{v} \Lambda_{mn}(X_{(j)}) \int_{X_{(j-1)}}^{X_{(j)}} \frac{1}{h_{mn}} K((x-t)/h_{mn}) dt , \quad 0 < x < T^*,$$

$$\tag{5.1.6}$$

where $v = \sup\{1 \le i \le r^*: X_{(i)} \le T^*\}$. Using (5.1.5) in the first integral of (5.1.6), expanding $\ln(1 - u)$ in series, and using the definition of $\hat{F}_{mn}(x)$, we can show that

$$\Lambda_{mn}^*(x) = \hat{F}_{mn}(x) + \sum_{j=2}^{\infty} (-1)^{j-1} \hat{F}_{mn}(x)[F_{mn}(x)]^{j-1} .$$

c) *Nadaraya-Type Estimator*: Denoted by $\hat{\Lambda}_{mn}(x)$, a "Nadaraya-type" estimator of $\Lambda(x)$ is defined as $\hat{\Lambda}_{mn}(x) = -\ln(1 - \hat{F}_{mn}(x))$, where $\hat{F}_{mn}(x)$ is the kernel estimator of F defined in (5.1.1). Again, expanding $\ln(1 - u)$ in series, we obtain

$$\hat{\Lambda}_{mn}(x) = \hat{F}_{mn}(x) + \sum_{j=1}^{\infty} j^{-1}(-1)^{j-1}[\hat{F}_{mn}(x)]^{j},$$

so this estimator is somewhat different than the kernel-type estimator in (b) above.

Finally, a smooth estimator of the hazard rate function λ is defined by

$$\lambda_{mn}(x) = \frac{d}{dx} \Lambda^{*}_{mn}(x)$$

$$= \int_{0}^{T^{*}} \frac{1}{h_{mn}^{2}} K'((x-t)/h_{mn}) \, \Lambda_{mn}(t)dt, \tag{5.1.7}$$

where the derivative is taken in the usual sense. As with the density estimator, it can be shown that $\lambda_{mn}(x)$ is analogous to the usual kernel-type estimator of the hazard rate function for i.i.d. samples.

5.2 Asymptotic Properties of the Smooth Estimators

All estimators defined in the equations (5.1.1) to (5.1.7) are consistent, with most of them satisfying strong uniform consistency. Similarly, asymptotic normality under standard conditions can be established for most of the above estimators. Before examining the asymptotic properties of the estimators, we state the following general conditions for h_{mn} and the kernel K, some or all of which are used to establish the asymptotic properties of the estimators.

(h.1) $h_{mn} \to 0$ as $m \to \infty$.

(h.2) $mh_{mn} \to \infty$ as $m \to \infty$.

(K.1) $\int |K(y)| \, dy < \infty$, $\sup |K(y)| < \infty$, and $|yK(y)| \to 0$ as $y \to \infty$.

(K.2) $K(y) \geq 0$ and $\int K(y) dy = 1$.

(K.3) $\int y K(y)\, dy = 0$ and $\int y^2 K(y)\, dy$ is nonzero and finite.

Since the properties of all the functions presented in (5.1.2) through (5.1.7) are dependent on the properties of $\hat{F}_{mn}(x)$, we start with the asymptotic results for $\hat{F}_{mn}(x)$.

Theorem 5.2.1: (*Strong Uniform Consistency of* $\hat{F}_{mn}$) Let F be a uniformly continuous function. Also, let the sequence h_{mn} satisfy (h.1) and K satisfy (K.1) and (K.2). Then as $m \to \infty$,

$$\sup_{x \geq 0} \left| \hat{F}_{mn}(x) - F(x) \right| \to 0 \quad \text{a.s.}$$

Proof: Define $F^0(x) = \int\limits_0^\infty K((x-t)/h_{mn})/h_{mn}\, F(t)\, dt$ and let

$$V_{mn} = \sup_{x \geq 0} \left| \hat{F}_{mn}(x) - F^0(x) \right|.$$

Now,

$$\sup_{x \geq 0} \left| \hat{F}_{mn}(x) - F(x) \right|$$

$$\leq V_{mn} + \sup_{x \geq 0} \left| F(x) - F^0(x) \right|. \qquad (5.2.1)$$

The second term on the right-hand side of (5.2.1) goes to zero by Nadaraya (1965). Hence, we only need to show that $V_{mn} \to 0$ with probability one as $m \to \infty$. But

$$V_{mn} = \sup_{x \geq 0} \left| \int\limits_0^\infty [F_{mn}(x) - F(x)] \frac{1}{h_{mn}} K((x-t)/h_{mn})\, dt \right|$$

$$\leq \sup_{x \geq 0} \left| F_{mn}(x) - F(x) \right| \left| \int\limits_0^\infty \frac{1}{h_{mn}} K((x-t)/h_{mn})\, dt \right|$$

$$\leq \sup_{x \geq 0} \left| F_{mn}(x) - F(x) \right| \to 0 \text{ a.s.}$$

by Samaniego and Whitaker (1988). ◆

Next, we look at the asymptotic normality of $\hat{F}_{mn}(x)$.

   Theorem 5.2.2: (*Asymptotic Normality of $\hat{F}_{mn}$*) In addition to the conditions of Theorem 5.2.1, assume (K.3) and that $m^{1/4}h_{mn} \to 0$ as $m \to \infty$. Then for all T such that $F(T) < 1$ and F'' is continuous on $[0,T]$, the process

$$\left\{ \hat{Y}_{mn}(x) = m^{1/2} \left[\hat{F}_{mn}(x) - F(x) \right]: \ 0 \le x \le T \right\}$$

has the same asymptotic distribution as the process

$$\left\{ Y_{mn}(x) = m^{1/2} \left[F_{mn}(x) - F(x) \right]: \ 0 \le x \le T \right\}.$$

   Proof: Recall from Chapter 4 that the process $\{Y_{mn}(x)\}$ is asymptotically a Gaussian process. Now write

$$\hat{Y}_{mn}(x) = I_1(x) + I_2(x) + Y_{mn}(x), \ 0 < x < T,$$

where

$$I_1(x) = \int_0^\infty \frac{1}{h_{mn}} K((x-t)/h_{mn}) \ \sqrt{m} \ \left[Z_m(t) - Z_m(x) \right] dt$$

with $Z_m(x) = F_{mn}(x) - F(x)$, and

$$I_2(x) = \int_0^\infty \frac{1}{h_{mn}} K((x-t)/h_{mn}) \ \sqrt{m} \ \left[F(t) - F(x) \right] dt.$$

Using a Taylor series expansion for $F(t) - F(x)$ in $I_2(x)$ gives

$$I_2(x) = \sqrt{m} \ f(x) \int_0^\infty \left[(x-t)/h_{mn} \right] K((x-t)/h_{mn}) \ dt$$

$$+ \frac{\sqrt{m}}{2} f'(x_0) \int_0^\infty \left[(x-t)/h_{mn} \right]^2 K((x-t)/h_{mn}) \ dt$$

for some x_0. So, by the conditions of the theorem, we have

$$|I_2(x)| \le \sqrt{m} \ h_{mn} \ f(x) \int_0^\infty y K(y) \ dy$$

$$+ \sqrt{m}\, h_{mn} [\sup f'(x)] \sup \int_0^\infty (y)^2 K(y)\, dy \to 0$$

as $m \to \infty$.

Next, for $\delta > 0$, define the interval $I_\delta = (x - \delta, x + \delta)$ at x. For m large enough, we have

$$|I_I(x)| \le \int_{I_\delta} h_{mn}^{-1} K((x-t)/h_{mn})\, \sqrt{m}\, |Z_m(t) - Z(t)|\, dt$$

$$\le \sup_{t \in I_\delta} \left\{ \sqrt{m}\, |Z_m(t) - Z_m(x)| \right\}.$$

Therefore, for $\varepsilon > 0$,

$$\lim_{m \to \infty} P\left(|I_1(x)| > \varepsilon \right) \le \lim_{m \to \infty} P\left(\sup_{t \in I_\delta} \left(\sqrt{m}\, |Z_m(t) - Z_m(x)| \right) \ge \varepsilon \right)$$

$$\le P\left(\sup_{t \in I_\delta} |Z_m(t) - Z_m(x)| \ge \varepsilon \right)$$

which goes to 0 as δ goes to 0 by the continuity properties of Gaussian processes. Therefore, for each x in $(0,T)$, $I_1(x) \xrightarrow{\ P\ } 0$ in probability and $I_2(x) \to 0$ as $m \to \infty$, giving the result. ◆

Most of the asymptotic properties of the other smooth estimators described in Section 5.2 follow directly from those of $\hat{F}_{mn}$. We present these here; some, which are easily proven, are stated without proof.

We now consider the asymptotic properties of the density estimator f_{mn}.

Theorem 5.2.3: (Strong Consistency of f_{mn}) Suppose that the bandwidth sequence h_{mn} satisfies (h.1) and the kernel K satisfies (K.1) and (K.2). In addition, also assume that (i) K is a function of bounded variation, (ii) K is symmetric with finite support, and (iii) f is uniformly continuous. Then as $m \to \infty$,

$$\sup_{x \ge 0} |f_{mn}(x) - f(x)| \to 0 \quad \text{a.s.}$$

Proof: We let $f^o(x) = \int_0^\infty h_{mn}^{-1} K((x-t)/h_{mn})f(t)dt$ and define $U_{mn} = \sup_{x > 0} |f_{mn}(x) - f^o(x)|$. Then by the triangle inequality, $\sup_{x > 0} |f_{mn}(x) - f(x)|$

$\leq U_{mn} + \sup_{x>0} |f^{\circ}(x) - f(x)|$. The second term on the right-hand side of this last inequality approaches zero as m goes to infinity by the results of Nadaraya (1965). Therefore, we need only show that U_{mn} goes to zero with probability one as m increases to infinity. To see this, note that

$$U_{mn} = \sup_{x \geq 0} \left| \int_0^\infty h_{mn}^{-1} K((x-t)/h_{mn}) d[F_{mn}(t) - F(t)] \right|$$

$$\leq \sup_{x \geq 0} \left| \int_0^\infty h_{mn}^{-1} [F_{mn}(t) - F(t)] dK((x-t)/h_{mn}) \right|,$$

yielding, by the properties of K and the results of Samaniego and Whitaker (1988), that

$$U_{mn} \leq \sup_{x \geq 0} \left(|F_{mn}(t) - F(t)| \cdot \left| \int_0^\infty h_{mn}^{-1} dK((x-t)/h_{mn}) \right| \right)$$

$$\leq M \sup_{x \geq 0} |F_{mn}(x) - F(x)| \to 0 \quad a.s.,$$

where M is the total variation of the kernel function, completing the proof. ◆

Theorem 5.2.4: (*Asymptotic Normality of f_{mn}*) Under the conditions of Theorem 5.2.2, for T such that $F(T) < 1$ and f'' is continuous on $[0,T]$, for each $0 \leq x \leq T$,

$$m^{1/2} \left[f_{mn}(x) - f(x) \right] \xrightarrow{d} Z(x),$$

where "d" denotes convergence in distribution and $Z(x)$ is a mean zero normal random variable with variance given by

$$\sigma^2 = \left(f^2(x) \int_{\overline{F}(x)}^1 \frac{1-u}{[u^2(1-u^n)]} du \right) + \left(\frac{f^2(x) F(x)}{\overline{F}(x)(1-\overline{F}^n(x))} \right).$$

Proof: For all $0 \leq x \leq T_F$, where $T_F = \sup\{y: F(y) < 1\}$, define

$$\hat{Y}_{mn}(x) = m^{1/2} \left\{ \hat{F}_{mn}(x) - F(x) \right\}.$$

Then from Theorem 5.2.2, $\lim_{m \to \infty} \hat{Y}_{mn}(x)$ is a second-order Gaussian process, say $Z(x)$. Next, define $Y_m(x) = \sqrt{m}\,(f_{mn}(x) - f(x)) = \hat{Y}'_{mn}(x)$, the derivative in the usual sense. Then, as m increases to infinity,

$$\int_0^t Y_m(x)dx = \hat{Y}_{mn}(t) \xrightarrow{\ d\ } Z(t).$$

Write

$$Y_m(x) = \hat{Y}'_{mn}(x) = \sqrt{m}\ [\int_0^{\infty} \frac{1}{h_{mn}} K((x-t)/h_{mn})\ dF_{mn}(t)$$

$$-\int_0^{\infty} \frac{1}{h_{mn}} K((x-t)/h_{mn})\ dF(t)$$

$$+\int_0^{\infty} \frac{1}{h_{mn}} K((x-t)/h_{mn})\ dF(t)$$

$$-\int_0^{\infty} \frac{1}{h_{mn}} K((x-t)/h_{mn})\ f(x)dt\ ]$$

$$= \sqrt{m}\ [\int_0^{\infty} \frac{1}{h_{mn}} K'((x-t)/h_{mn})\ (F_{mn}(t) - F(t))\ dt$$

$$+\int_0^{\infty} \frac{1}{h_{mn}} K((x-t)/h_{mn})\ (f(t) - f(x))\ dt\ ]. \qquad (5.2.2)$$

By Taylor's expansion, the second term in (5.2.2) can be represented as

$$\sqrt{m}\ h_{mn}\ f'(x)\int_0^{\infty} y\,K(y)dy + \frac{\sqrt{m}}{2}\ h_{mn}^2\ f''(x_0) \int y^2 K(y)\ dy$$

for some x_0. Thus,

$$Y_m(x) = \sqrt{m}\ [\int_0^{\infty} \frac{1}{h_{mn}} K'((x-t)/h_{mn})\ (F_{mn}(t) - F(t))\ dt\ +$$

$$\sqrt{m}\ C(x_0)\,h_{mn}^2\ ,$$

where $C(x_0)$ is the constant coefficient in the previous expression, and for m large,

$$Y_m(x) \approx \int_0^\infty \frac{1}{h_{mn}^2} K'((x-t)/h_{mn}) Z(t)\, dt = Z_0'(t)$$

is a mean zero Gaussian process. Therefore,

$$\int_0^t Y_m(x)dx \approx \int_0^t Z_0'(x)dx = Z_0(t),$$

which has the same distribution as $Z(t)$ by Hoel, Port, and Stone (1972). Then $Z_0(t)$ is a mean zero process with covariance function given by

$$r(s, t) = \frac{\partial^2}{\partial s \partial t} R(s,t),$$

where $R(s,t)$ is the covariance function for $Z(t)$. ♦

From here, we consider the properties of the quantile estimators, starting with the empirical quantile estimator. The strong uniform consistency of Q_{mn} can be established by using arguments similar to those used by Csörgő (1983, Chapter I, pp. 1 – 10), to prove consistency for the quantile functions obtained from complete data. The following theorem is stated without proof.

Theorem 5.2.5: (Strong Uniform Consistency of Q_{mn}) Assume that F has finite support and is a twice-differentiable function on $(0, \infty)$. Assume that $\inf_{0 \le p \le 1}(f(F^{-1}(p))) > 0$ and $\sup_{0 \le p \le 1} |f'(F^{-1}(p))| < \infty$. Then

$$\sup_{0 < p < 1} |Q_{mn}(p) - Q(p)| \to 0 \text{ a.s. as } m \to \infty.$$

We now establish that the quantile process,

$$\rho_m(p) = m^{1/2}(f(Q(p)))[Q_{mn}(p) - Q(p)]$$

converges weakly to a mean zero Gaussian process. The argument used is the same as the one used by Shorack and Wellner (1986) to prove the convergence of Kaplan–Meier quantiles. The proof for the asymptotic normality of the empirical quantile estimator is not as straightforward and is presented in detail after the following preliminary definitions.

Let $(D[a,b], \|.\|) = (D, \|.\|)$ be the metric space of real-valued right-continuous functions on $[a, b]$ that have finite left-hand limits using supnorm as the metric. Let $(D_0, \|.\|)$ be the metric space of nonnegative nondecreasing functions in $(D, \|.\|)$ and let $(C, \|.\|)$ be the metric space of continuous functions on $[a, b]$. Recall that from Chapter 4, $F_{mn} \in D_0$ and $F \in C$ with $a = 0$ and $b = T_F$.

The following lemma of Vervaat (1972) is a key part of the proof of the asymptotic normality of Q_{mn}.

Lemma 5.2.1: Let $Y_n(t)$ be a stochastic process in D_0, let Y be a random element in C, and let $\{\varepsilon_n\}_1^\infty$ be a divergent sequence of positive real numbers. Then $\varepsilon_n (Y_n(t) - t)$ converges weakly to Y if and only if $\varepsilon_n (Y_n^{-1}(t) - t)$ converges weakly to $-Y$.

This brings us to the asymptotic normality of $\rho_m(p)$.

Theorem 5.2.6: (Asymptotic Normality of Q_{mn}) Let F be absolutely continuous with density f. Assume that $f(F^{-1}(\cdot))$ is bounded away from zero and continuous on $[0, F(T_F)]$. Then the process

$$\{\rho_m(p) = m^{1/2} (f(Q(p)))[Q_{mn}(p) - Q(p)]\}$$

converges weakly to $-Z(F^{-1}(\cdot))$ on $D[0, F(T_F)]$, where $Z(\cdot)$ is a mean zero Gaussian process with covariance function given by $R(s, t)$.

Proof: Again from Chapter 4, the process

$$Z_m(F^{-1}(p)) = \sqrt{m}\left[F(F_{mn}^{-1}(p)) - p\right]$$

converges weakly to $-Z(F^{-1}(.))$ on $D[0, F(T_F)]$, where $Z(\cdot)$ is a mean zero Gaussian process with covariance function $R(s, t)$. Now define the process

$$\rho_m^*(p) = \sqrt{m}\left[F(F_{mn}^{-1}(p)) - p\right].$$

By Lemma 5.2.1, the process $\rho_m^*(p)$ converges weakly to $-Z(F^{-1}(p))$ on $D[0, F(T_F)]$. Next note that

$$r_m(p) = \sqrt{m} \ (f(Q(p)))\left[F_{mn}^{-1}(p) - F^{-1}(p)\right]$$

$$= \frac{mf(Q(p)) \left[Q_{mn}(p) - Q(p)\right] \rho_m^*(p)}{F(F_{mn}^{-1}(p)) - p}. \tag{5.2.3}$$

Using the approach of Csörgő (1983), the right-hand side of (5.2.3) is

$$\frac{\rho_m^*(p)\,f(F^{-1}(p))}{f(F^{-1}(p_n^*))},$$

where $\left| p - p_n^* \right| < \left| F(\,F_{mn}^{-1}(p)) - p \right|$. Therefore,

$$\sup_{0 \le p \le F^{-1}(T_F)} \left| \rho_m(p) - \rho_m^*(p) \right| \le \sup_{0 \le p \le F^{-1}(T_F)} \left| \rho_m(p) \right| \left| \frac{f(F^{-1}(p))}{f(F^{-1}(p_n^*))} - 1 \right|. \quad (5.2.4)$$

But, the right-hand side of (5.2.4) approaches 0 as $m \to \infty$ from the weak convergence of $\rho_m^*(p)$ and the convergence of $\left(F(F_{mn}^{-1}(p)) - p\right)$ to 0 (see Csörgő, 1983). Thus ρ_m^* and ρ_m have the same asymptotic distribution. ♦

The asymptotic properties of $\hat{Q}_{mn}$ follow directly from those of Q_{mn}, and the next two theorems give these results. The proofs follow fairly easily from the results in Theorems 5.2.1 and 5.2.2.

Theorem 5.2.7: (*Strong Uniform Consistency of* $\hat{Q}_{mn}$) Let F be uniformly continuous satisfying the conditions for strong uniform consistency of $\hat{F}_{mn}$. Let the sequence h_{mn} satisfy (h.1) and the kernel K satisfy (K.1) and (K.2). Then as $m \to \infty$,

$$\sup_{0<p<1} |\hat{Q}_{mn}(p) - Q(p)| \to 0 \quad \text{a.s.}$$

Proof: First, we define

$$Q_0(p) = \int_0^1 h_{mn}^{-1} K((p - t)/h_{mn})Q(t)dt$$

and let

$$W_{mn} = \sup_{0<p<1} |\hat{Q}_{mn}(p) - Q^0(p)|.$$

Then $\sup_{0<p<1} |\hat{Q}_{mn}(p) - Q(p)| \le W_{mn} + \sup_{0<p<1} |Q^0(p) - Q(p)|$. The second term on the right-hand side of the inequality approaches zero as m increases to infinity by results of Nadaraya (1965), and W_{mn} goes to zero a.s. as m increases from the conditions on K and the strong consistency of Q_{mn}. ♦

Theorem 5.2.8: (*Asymptotic Normality of* $\hat{Q}_{mn}$) Assume that the conditions for consistency of $\hat{Q}_{mn}$ and asymptotic normality of Q_{mn} and

$\hat{F}_{mn}$ are satisfied. Then for a fixed $0 < p < 1$, the random variable $\hat{\rho}_m (p) = m^{1/2} f(Q(p)) \left[\hat{Q}_{mn} (p) - Q(p) \right]$ converges in distribution to $Z(p)$, where $Z(p)$ is a mean zero normal random variable with variance given by

$$\sigma_p^2 = (1 - p)^2 \int_{1-p}^{1} \left(u \sum_{j=1}^{n} u^j \right)^{-1} du.$$

Proof: The proof of this theorem follows similarly to that of Theorem 5.2.2. For large enough m, write $\hat{\rho}_m (p) = I_1 - I_2 + I_3 + \rho_m(p)$, where

$$I_1 = m^{1/2} f(Q(p)) \int_0^1 [Q_{mn}(t) - Q(t)] h_{mn}^{-1} K((p - t)/h_{mn}) dt,$$

$$I_2 = m^{1/2} f(Q(p)) \int_0^1 [Q_{mn}(p) - Q(p)] h_{mn}^{-1} K((p - t)/h_{mn}) dt,$$

and

$$I_3 = m^{1/2} f(Q(p)) \int_0^1 [Q_{mn}(t) - Q(p)] h_{mn}^{-1} K((p - t)/h_{mn}) dt.$$

Then it can be shown that I_1 and I_2 converge to zero in probability as m increases, and that I_3 converges to zero by Yang (1985). Thus, by Slutsky's theorem, $\hat{\rho}_m (p)$ and $\rho_m(p)$ have the same asymptotic distribution, completing the proof. ♦

Using the strong uniform consistency of $\hat{F}_{mn}$ and the uniform continuity of F gives consistency of the Nadaraya-type estimator of $Q(p)$. This is stated next.

Theorem 5.2.9: (*Strong Pointwise Consistency of x_{mn}*) Let F be uniformly continuous and strictly increasing. Let the sequence h_{mn} satisfy (h.1) and the kernel K satisfy (K.1) and (K.2). Then, for a given $0 < p < 1$, $x_{mn}(p) \to Q(p)$ a.s. as $m \to \infty$.

Using a Taylor series expansion for $\hat{F}_{mn} (x_{mn}(p)) - \hat{F}_{mn}(Q(p))$ and the normality of $\hat{F}_{mn}$ gives the asymptotic normality for x_{mn}, stated next.

Theorem 5.2.10: (Asymptotic Normality of x_{mn}) Assume the conditions for consistency of x_{mn} and that $Q(p) < T_F$. In addition, assume that the kernel function K satisfies (K.3) and that $m^{1/4} h_{mn} \to 0$ as $m \to \infty$. Then for a fixed $0 < p < 1$, the random variable

$$Z_{mn}(p) = m^{1/2} f(Q(p))[x_{mn}(p) - Q(p)]$$

converges in distribution to a mean zero normal random variable with variance given by σ_p^2, where σ_p^2 was defined in Theorem 5.2.8.

Proof: By a Taylor series expansion,

$$\hat{F}_{mn}(x_{mn}(p)) - \hat{F}_{mn}(Q(p)) = [x_{mn}(p) - Q(p)]\hat{f}_{mn}(p^*),$$

where p^* is a random point between $x_{mn}(p)$ and $Q(p)$ and $\hat{f}_{mn}$ is the derivative of $\hat{F}_{mn}$.
Then

$$Z_{mn}(p) = m^{1/2} f(Q(p))[x_{mn}(p) - Q(p)]$$

$$= -m^{1/2}[\hat{F}_{mn}(Q(p)) - F(Q(p))]\left[\frac{f(Q(p))}{\hat{f}_{mn}(p^*)}\right],$$

which by Slutsky's theorem has the same asymptotic distribution as $m^{1/2}[\hat{F}_{mn}(Q(p)) - F(Q(p))]$, since $[f(Q(p))/\hat{f}_{mn}(p^*)]$ converges almost surely to one from the results of Gulati and Padgett (1994a). ◆

Finally, we consider the properties of the smooth estimators of the hazard and the hazard rate functions. The following results are stated without proof since the consistency, and the asymptotic normality of these estimators follows from the same results either for F_{mn} or for the corresponding discrete estimators. We start with the asymptotic normality for Λ_{mn}, the discrete estimator of the hazard function defined in Chapter 4.

Theorem 5.2.11: (Asymptotic Normality of Λ_{mn}) For all T such that $F(T) < 1$, the process $\left\{ m^{1/2}(\Lambda_{mn}(t) - \Lambda(t)) : 0 \le t \le T \right\}$ converges weakly to a mean zero Gaussian process with covariance function

$$r(s,t) = \int\limits_{0}^{s \wedge t} \frac{1}{E(\kappa(t))} \, d\Lambda(x),$$

where $s \wedge t = \min(s,t)$ and $\kappa(x)$ was defined in Chapter 4.

The above theorem is proven using the fact that $\Lambda_{mn}(x)$ is a continuous differentiable function of an asymptotic Gaussian process F_{mn} and finite-dimensional (multivariate) normality arguments (see Gulati and Padgett, 1994c).

The strong uniform consistency and normality of $\Lambda_{mn}^{*}(x)$ then follow under standard conditions on the bandwidth and the kernel function. The statements of the theorems and the proofs are similar to the ones stated above in this chapter and are not repeated here. Similarly, $\hat{\Lambda}_{mn}(x)$ is shown to be asymptotically normal and strongly consistent. Gulati and Padgett (1994c) have been able to establish strong pointwise consistency for λ_{mn}, but not normality or strong uniform consistency.

Bias or mean-square error results for these estimators are not available at present. Neither are any results on the rates of convergence for the various estimators. Moreover, with the lack of an expression for the MSE of the smooth estimators, an optimal value of h cannot be evaluated. Therefore Gulati and Padgett (1994(a,b,c)) investigated the small sample properties of the smooth estimators through computer simulations. The purpose of these simulations was to examine the effect of the bandwidth and the sample sizes on the biases and the mean-squared errors of the estimators. They considered various kernels, such as the triangular, uniform, and truncated normal, and various underlying distributions, including exponential, Weibull, normal, and log-normal. In all cases investigated, they found that for each function, there was at least one value of h that locally minimized the estimated mean-squared error. This minimizing value of h increases in the right tail of every distribution considered, as would be expected from the nature of the data, since observed minima are sparse in the right tail. It was also observed that in all cases considered, the biases seemed to follow the pattern of the mean-squared errors.

Table 5.1 shows some simulated mean-square errors for $\hat{F}_{mn}$, the smooth estimator of the distribution function, and Table 5.2 compares the mean-squared errors of the quantile estimator Q_{mn} with the estimators $\hat{Q}_{mn}$ and x_{mn}.

Table 5.1. Simulated Mean-Squared Errors for $\hat{F}_{mn}$ for the Standard Lognormal Distribution, Truncated N(0,1) Kernel, $m = 30$, $n = 20$

h \ x	0.51	1.51	2.01
0.11	0.658	3.223	3.823
0.16	0.595	3.049	3.650
0.21	0.545	2.869	3.466
0.26	0.502	2.689	3.281
0.31	0.465	2.525	3.100
0.36	*0.446	2.392	2.928
0.41	0.454	2.311	2.772
0.46	0.498	*2.299	2.638
0.51	0.583	2.371	2.537
0.56	0.711	2.536	2.478
0.61	0.881	2.794	*2.473
0.66	1.092	3.141	2.533
0.71	1.340	3.569	2.666
0.76	1.622	4.065	2.878
0.81	1.935	4.617	3.172
0.86	2.276	5.211	3.549

(Minimum MSE values are indicated by *.)
(All entries are multiplied by 10^3.)

As an example, Gulati and Padgett also computed the values of the estimators for the record values obtained from successive failure times (in hours) of air conditioning units on Boeing 720 aircraft introduced by Proschan (1963). These data were discussed in Chapter 3, where it was mentioned that Proschan tested and accepted the hypothesis that the successive failure times were i.i.d. exponential. The estimators were calculated for h values between 0.09 and 2.01 and the graph of the function that was "most pleasing to the eye" was picked as the final estimate. In all cases considered, they found that the smooth estimators calculated from the records were fairly good approximations of the corresponding functions from the exponential distribution. Figure 5.1 gives the graph of the estimated smooth density function for the air conditioner data for the normal kernel with the "most pleasing" value of the smoothing parameter. In the same vein, Figure 5.2 gives the graph of the "most pleasing" smoothed hazard rate function for the air conditioning units.

Table 5.2. MSE Comparison of Q_{mn} with $\hat{Q}_{mn}$ and x_{mn}. for the Standard
Exponential Distribution, Triangular Kernel,
$m = 60$, $n = 30$

h \ p	Ratio 1				Ratio 2		
	0.1	0.5	0.75		0.1	0.5	0.75
0.09	1.401	1.046	1.050		*1.377	1.060	1.150
0.14	*1.465	1.070	1.068		1.350	1.110	1.115
0.19	1.335	1.102	1.094		0.861	*1.126	0.998
0.24	1.003	1.142	1.121		0.412	1.086	0.779
0.29	0.562	1.184	1.145		0.201	0.973	0.658
0.34	0.282	1.222	1.165		0.109	0.803	0.887
0.39	0.148	1.247	1.179		0.065	0.615	1.407
0.44	0.084	*1.258	1.189		0.042	0.441	*1.819
0.49	0.051	1.252	1.196		0.028	0.294	1.430
0.54	0.033	1.226	1.200		0.020	0.208	0.885
0.59	0.023	1.181	*1.201		0.014	0.191	0.559
0.64	0.017	1.119	1.198		0.011	0.202	0.381
0.69	0.013	1.043	1.190		0.0084	0.234	0.278
0.74	0.012	0.958	1.178		0.0065	0.290	0.215
0.79	0.0117	0.867	1.161		0.0050	0.382	0.173
0.84	0.0115	0.777	1.138		0.0039	0.527	0.143

Ratio 1 = MSE(Q_{mn})/MSE(x_{mn}).
Ratio2 = MSE(Q_{mn})/MSE ($\hat{Q}_{mn}$).
(Minimum MSE values for the smooth estimators are indicated by *.)

The results on smooth estimation from randomly sampled record-breaking data were extended to inversely sampled data by Gulati and Padgett (1995). The stochastic form of the functions and their asymptotic properties remains the same for the inversely sampled case and therefore, is not presented here. As far as small sample properties are concerned, Gulati and Padgett (1995) decided to compare the performance of the nonparametric estimators under the two sampling schemes. Their simulation studies, however, did not show a signif-icant difference in the performance of the smooth estimators under the two sampling schemes. In their simulation study, Gulati and

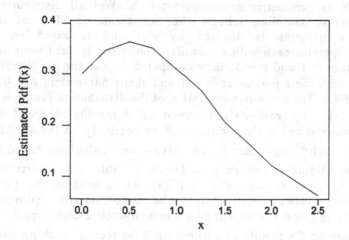

FIGURE 5.1. Kernel Density Estimator for Air Conditioner Data Using Normal Kernel

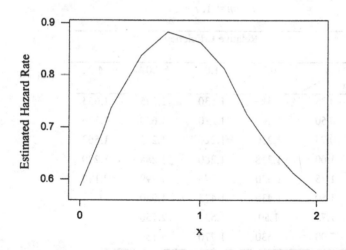

FIGURE 5.2. Smooth Estimate of Hazard Rate Function for Air Conditioning Units

Padgett (1995) also included a comparison of the nonparametric empirical estimator of the distribution function for the inverse sampling scheme with its parametric counterpart for the Weibull distribution. (For the inverse sampling scheme the nonparametric MLE of the underlying distribution is denoted by $F_{m,r}$ and is based on m independent samples each with r records.) Using the inverse sampling scheme with $m = 1$ and $r = 5$, they computed $F_{1,5}(x)$ for the Weibull distribution with scale parameter $\alpha = 1$ and shape parameters $\beta = 0.8$, 1.0, 2.0, and 4.0. The corresponding MLE of the distribution function is given by $F_P(x) = 1 - \exp(-ax^b)$, where a and b are the MLEs of the shape parameter α and scale parameter β, respectively, in the Weibull density, $f(x) = \alpha\beta x^{\beta-1} \exp(-\alpha x^\beta)$. (The MLEs were calculated based on the paper by Hoinkes and Padgett (1994).) Table 5.3 reports the efficiency of $F_{1,5}(x)$ as compared to $F_P(x)$. As is evident, the parametric estimator always outperforms the nonparametric estimator, sometimes being twice as efficient as the nonparametric counterpart.

On examining the results on estimation from record-breaking data, one can thus conclude that nonparametric modeling from this type of data works well. In fact, every possible attribute of the underlying distribution was examined under the record-breaking setting and it was

Table 5.3. MSE Comparison of $F_{1,5}(x)$ with $F_P(x)$ for the Weibull Distribution, $m = 1, r = 5$

	Relative Efficiency			
x \ β	0.8	1.0	2.0	4.0
0.25	1.513	1.630	1.225	1.653
0.50	1.240	1.280	1.630	1.226
0.75	1.230	1.220	1.263	1.562
1.00	1.288	1.280	1.288	1.289
1.25	1.330	1.378	1.499	2.120
1.50	1.430	1.473	1.960	
1.75	1.50	1.537	2.750	
2.00	1.630	1.770	4.130	

Relative Efficiency = MSE($F_{1,5}$)/MSE(F_P).

found that all the estimators (smooth and discrete) for these functions had good global properties. Yet another approach to modeling from record-breaking data is Bayesian inference. This approach is reviewed in the next chapter.

6
Bayesian Models

As seen in the previous three chapters, parametric inference from record-breaking data can be quite challenging, and nonparametric inference is perhaps even more so. Even for the case of complete random samples, Bayesian inference can be quite complex and require highly sophisticated computational methods, such as Markov chain Monte Carlo, to obtain posterior distributions. It is not surprising then that very little research has been done on Bayesian estimation from record-breaking data. Initial research from this perspective can be found in Dunsmore (1983) who provided Bayesian predictive distributions for $X_m - X_n$ for an exponential distribution (both the one-parameter and the two-parameter models). This was followed by the work of Basak and Bagchi (1990) who developed an approximation for the predictive distribution of a future record using past records, based on Laplace approximation. Tiwari and Zalkikar (1991) considered the general problem of nonparametric Bayesian inference from record-breaking data. They derived the nonparametric Bayes and the empirical Bayes estimators of the underlying survival function for such data under a Dirichlet process prior and squared-error loss function. In this chapter, all of the work done on Bayesian inference from record-breaking data is summarized, starting with the work of Dunsmore (1983).

6.1 Bayesian Prediction Intervals

Dunsmore (1983) considers a Bayesian approach for the case where one observes upper records along with the corresponding record times. Again assume that $Y_1, Y_2, \ldots,$ is a random sample from a continuous distribution function F with density f. Thus the observed data consist of the sequence $X_{N_1}, X_{N_2}, \ldots,$ where $X_{N_i}, i = 1, 2, \ldots,$ is the ith new maximum and $N_i, i = 1, 2, \ldots,$ is the time at which the ith new maximum is observed, that is, $N_k = \min\{j: j > N_{k-1}, Y_j > X_{N_{k-1}}\}$. Assuming that one has the first m records available from an underlying exponential distribution, Dunsmore provides Bayesian prediction intervals for the difference $X_{N_{m+r}} - X_{N_m}$. The approach is similar for both the one-parameter and the two-parameter exponential model. Suppose that the data can be summarized by a sufficient statistic Z for the unknown parameter θ with the underlying model $f(z|\theta)$. Further let $p(\theta)$ be the given prior on the parameter space Θ. The prior can then be updated to a posterior $f(\theta|z)$ based on the data, and information about a future value y can be obtained from the predictive density function

$$f(y|z) = \int_\theta f(y|\theta) f(\theta|z) \, d\theta.$$

We first consider the case of the one-parameter exponential distribution.

6.1.1 One-Parameter Exponential Model

Here the underlying density is $f(z|\theta) = \theta \exp(-\theta z)$, $z > 0$. From Dunsmore (1983), $Z = X_{N_m}$ is a sufficient statistic for θ. If we take the prior $p(\theta)$ to be the conjugate gamma with shape parameter g and scale parameter h, then the predictive distribution of the increase $y = X_{N_{m+r}} - X_{N_m}$ is given by

$$p(y|z) = \frac{H^G \, y^{r-1}}{B(r, G) \, (y + H)^{G+r}}, \quad y > 0, \tag{6.1.1}$$

where $G = g + m$, $H = h + X_{N_m}$, and $B(a,b)$ is the complete beta function given by

$$B(a,b) = \frac{\Gamma(a) \, \Gamma(b)}{\Gamma(a+b)}.$$

The distribution in (6.1.1) is henceforth denoted by IN Be(r, G, H), also called the beta density of the second type (see Arnold et al., 1998, p. 162).

6.1.2 Two-Parameter Exponential Model

Here the underlying density is $f(z|\theta) = \tau \exp(-\tau(z - \mu))$, $z > \mu$. Again, from Dunsmore (1983), the statistics X_{N_1} and $X_{N_m} - X_{N_1}$ are jointly sufficient for $\theta = (\tau, \mu)$. Let the prior $p(\theta)$ be the conjugate exponential-gamma distribution, ElGa (b, c, g, h) for θ; that is,

$$p(\tau, \mu) = c\,\tau \exp\{-c\tau(b - \mu)\}\, \frac{h^g\, \tau^{g-1}\, \exp\,(-h\tau)}{\Gamma\,(g)}, \quad \mu < b, \tau > 0.$$

Once again, the predictive distribution of the increase $y = X_{N_{m+r}} - X_{N_m}$ is given by a beta density of the second type, specifically, the IN Be(r, G, H) distribution, where now

$$G = g + m - 1 + \delta(c),$$

$$H = \begin{cases} h + X_{N_m} - X_{N_1} - c(X_{N_1} - b), & X_{N_1} < b, \\ h + X_{N_m} - X_{N_1} + (X_{N_1} - b), & X_{N_1} \geq b, \end{cases}$$

and

$$\delta\,(c) = \begin{cases} 0 & \text{if } c = 0, \\ 1 & \text{if } c > 0. \end{cases}$$

6.2 Laplace Approximations for Prediction

Based on the Laplace approximation (as developed by Tierney and Kadane, 1986), Basak and Bagchi (1990) developed an approximation for the predictive distribution of a future record using past records. As in the previous section, assume again that $Y_1, Y_2, \ldots$ is a random sample from a continuous distribution with density $f(y \mid \theta)$. Let the observed data consist of the sequence of m upper records, $R_m = \{ X_{N_1} = x_1, X_{N_2} = x_2, \ldots, X_{N_m} = x_m \}$. Furthermore, let $p(\theta)$ be the prior distribution defined on the parameter space Θ. It is well known that the joint distribution of $x_1, x_2, \ldots, x_m$ is given by

$$f(x_1, x_2, \ldots, x_m \mid \theta) = f(x_m \mid \theta) \prod_{i=1}^{m-1} \frac{f(x_i \mid \theta)}{1 - F(x_i \mid \theta)}, \quad x_1 < x_2 < \cdots < x_m.$$

$$(6.2.1)$$

By Bayes' theorem, the posterior distribution of θ is given by

$$p(\theta \mid R_m) = \frac{f(x_m|\theta) \prod_{i=1}^{m-1} r(x_i|\theta) \, p(\theta)}{\int_0^\infty f(x_m|\theta) \prod_{i=1}^{m-1} r(x_i|\theta) \, p(\theta) \, d\theta}, \tag{6.2.2}$$

where

$$r(.\mid\theta) = \frac{f(.\mid\theta)}{1 - F(.\mid\theta)}.$$

Hence, if X_{m+1} is the $(m+1)$st record from the same model, then the predictive density of X_{m+1} based on the records is given by the density

$$g(x_{m+1} \mid R_m) = \frac{\int f(x_{m+1}|\theta) \prod_{i=1}^{m} r(x_i|\theta) \, p(\theta) \, d\theta}{\int_\Theta f(x_m|\theta) \prod_{i=1}^{m} r(x_i|\theta) \, p(\theta) \, d\theta}. \tag{6.2.3}$$

The density (6.2.3) is analytically intractable whenever the product hazard function cannot be written in terms of simple functions. One could either use computationally intensive numerical integration techniques or resort to approximations. Basak and Bagchi propose using Laplace approximations to obtain the density of the future record as outlined next.

Let $L(\theta|D_m)$ be the likelihood of θ based on the m observations denoted by the set $D_m = (X_1, X_2, \ldots, X_m)$. The observations are not necessarily independent or identically distributed. Let $p(\theta)$ be the prior distribution defined on the parameter space Θ. Let $p(\theta|D_m)$ denote the posterior distribution of θ. Let X be a future observation generated from the same system.

The predictive distribution of X, given the data, is

$$h(x \mid D_m) = \int_\Theta f(x|\theta) \, p(\theta \mid D_m) \, d\theta$$

when the observations are i.i.d., and

$$h(x \mid D_m) = \frac{\int_\Theta L(\theta \mid D_{m+1}) \, p(\theta) \, d\theta}{\int_\Theta L(\theta \mid D_m) \, p(\theta) \, d\theta}$$

when the observations are not independent. In either case, define $L^*(\theta)$ to be the logarithm of the integrand in the numerator divided by m and let $L(\theta)$ denote the logarithm of the integrand in the denominator divided by m. Then

$$h(x|\ D_m) = \frac{\int\limits_{\Theta} e^{mL^*(\theta)}\ d\theta}{\int\limits_{\Theta} e^{mL(\theta)}\ d\theta} \ . \qquad (6.2.4)$$

Using the Laplace approximation simultaneously on the numerator and the denominator, the approximate predictive density is

$$\hat{h}(x|D_m) = \left(\frac{\det \Sigma^*}{\det \Sigma}\right)^{1/2} \exp\left\{ m(L^*(\hat{\theta}^*) - L(\hat{\theta})) \right\}, \qquad (6.2.5)$$

where $\hat{\theta}^*$ is the mode of L^*, $\hat{\theta}$ is the mode of L, and Σ^* and Σ are minus the inverse Hessians of L^* and L at $\hat{\theta}^*$ and $\hat{\theta}$, respectively. Now let D_m denote a set of m records from a distribution with density $f(y|\theta)$. Let $p(\theta)$ be the prior distribution defined on the parameter space Θ and let the parameter space be the entire real line. Suppose now that we are interested in predicting a future record (note that this is actually the only logical situation in many instances; for example, in marathons, we normally only have records and not the entire set of timings). Hence, we want to find the approximate predictive distribution of the next record, $X_{(m+1)} = x_{m+1}$. Comparing equations (6.2.3) and (6.2.4), we have that

$$mL(\theta) = \sum_{i=1}^{m-1} \log r(x_i|\theta) + \log f(x_m|\theta) + \log p(\theta),$$

and

$$mL^*(\theta) = \log r(x_m|\theta) + \log f(x_{m+1}|\theta) + mL(\theta),$$

where $r(\ .|\ \theta)$ is the hazard function defined earlier.

Suppose now that $\hat{\theta}^*$ is the mode of L^* and $\hat{\theta}$ is the mode of L, and let

$$\sigma^{*2} = -\frac{1}{mL^{*\prime\prime}(\hat{\theta}^*)} \quad \text{and} \quad \sigma^2 = -\frac{1}{mL''(\hat{\theta})} \ .$$

Then from equation (6.2.5), an approximate predictive distribution of $X_{(m+1)}$ (normalized by a constant $C(D_m)$) is

$$\hat{f}_1 (x_{m+1}|D_m) = \frac{\sigma^*}{\sigma} \exp\left[m(L^*(\hat{\theta}^*) + \frac{1}{m} \log \sigma^*)\right]. \qquad (6.2.6)$$

Basak and Bagchi (1990) found that for tractable distributions such as the exponential, the one-parameter Weibull, and the one-parameter Pareto, the approximate distribution of the $(m + 1)$st record is the same as its exact distribution. For other distributions, they show that the approximate predictive distribution of the future record values is easily computable with sufficient accuracy. Equation (6.2.6) can also be used to test if a recorded observation is a record by calculating the probability of observing a record at least as large as the one observed.

Example 6.2.1. (Basak and Bagchi, 1990)
Let $D_m = (X_1, X_2, \ldots, X_m)$ be the set of m records from the exponential distribution with parameter θ. Let X be an observation from the same system independent of the past (not necessarily a record). The density of X then is

$$f(y| \theta) = \theta \exp(-\theta x), \quad \theta > 0, x > 0.$$

Suppose *a priori* that θ has the gamma distribution with parameters g and h. Then the exact predictive distribution of X, given the data, is

$$h(x| D_m) = \frac{(m + g) (h + x_m)^{m+g}}{(h + x_m + x)^{m+g+1}},$$

and the probability that X is a future record is

$$P(X > x_m| D_m) = \left(\frac{h + x_m}{h + 2x_m} \right)^{m+g}.$$

For the Laplace approximation, we have

$$\hat{\theta}^* = \frac{m + g}{h + x_m + x} \quad \text{and} \quad \hat{\theta} = \frac{m + g - 1}{h + x_m},$$

and so as a result,

$$\sigma^{*2} = \frac{\hat{\theta}^{*2}}{m+g} = \frac{m+g}{h+x_m+x}.$$

Similarly,

$$\sigma^2 = \frac{m+g-1}{(h+x_m)^2},$$

and the predictive density of the future record is

$$\hat{f}_1(x_{m+1}|D_m) = e^{-1}\frac{(m+g)^{m+g+1/2}}{(m+g-1)^{m+g-1/2}} \cdot \frac{(h+x_m)^{m+g}}{(h+x_m+x)^{m+g+1}}.$$

6.3 Bayesian Inference for the Survival Curve

Tiwari and Zalkikar (1991) were the first to consider the general problem of Bayesian inference from record-breaking data. They derived the nonparametric Bayes and empirical Bayes estimators for record-breaking data under a Dirichlet prior process and squared-error loss function. As in the case of Samaniego and Whitaker (1988), they first examined the problem for a single record-breaking sample.

Let P be the probability measure associated with the underlying distribution function F, and assume that P has a Dirichlet process prior D_α on $(R^+, \beta(R^+))$ with parameter α, where $\beta(R^+)$ is the Borel field of subsets of the positive real line R^+. Let the observed lower record-breaking data be denoted by $X_1, K_1, X_2, K_2, \ldots, X_r, K_r$.

For notational ease, Tiwari and Zalkikar represent the data as $\{(Z_i, \delta_i), i = 1, 2, \ldots, n\}$, where $Z_i = \min\{Z_{i-1}, Y_i\}$ and $\delta_i = I[Y_i < Z_{i-1}]$ and $Z_1 = Y_1$.

Note that the pairs (Z_i, δ_i) are neither independent nor identically distributed. However, Z_{i-1} and Y_i are independent for each $i = 1, 2, \ldots, n$. Also note that when $\delta_i = 1$, then Z_i corresponds to a record value X_i and the number of δs equal to zero between two consecutive δ values that equal one corresponds to the values of K, that is, the number of observations needed to obtain a new record.

Using the above notation, Tiwari and Zalkikar (1991) show that the Bayes estimator $\hat{\bar{F}}_\alpha(t)$ of the survival function $\bar{F}(t)$, under squared-error loss and a Dirichlet process prior, is given by

$$\hat{\bar{F}}_\alpha(t) = B_n(t) \, W_n(t), \qquad t > 0, \tag{6.3.1}$$

where

$$B_n(t) = \frac{\alpha(t) + N_n^+(t)}{\alpha(R^+) + n}$$

and

$$W_n(t) = \prod_{j=1}^{n} \left(\frac{\alpha(Z_j) + N_n^+(Z_j) + \lambda_j}{\alpha(Z_j) + N_n^+(Z_j)} \right)^{I\left(Z_j \leq t, \, \delta_j = 0\right) / \lambda_j}$$

with the notation that $\alpha(u) = \alpha(u, \infty)$, $N_n^+(t)$ is the total number of Z_js greater than t, λ_j is the number of values equal to Z_j, and $I(\cdot, \cdot)$ is the indicator function.

Note that the case when $\alpha(R^+) \to 0$ corresponds to the prior being noninformative. In this case, it can be shown that the Bayes estimator defined in (6.3.1) reduces to the nonparametric maximum likelihood estimator

$$\bar{F}^*(t) = \prod_{\{i: \, X_{(i)} \leq t\}} \frac{\sum_{j=i}^{r} K_{(j)} - 1}{\sum_{j=i}^{r} K_{(j)}}$$

which was defined in (4.3.5) as the NPMLE of the survival function developed by Samaniego and Whitaker (1988).

Example 6.3.1. (Tiwari and Zalkikar, 1991)
To illustrate the calculation of the Bayes survival function for record-breaking data, Tiwari and Zalkikar (1991) consider Proschan's (1963) Boeing 720 air-conditioner data for plane 7912. The data presented in Table 6.3.1 were modified to suit the setup of their calculations. Tiwari and Zalkikar consider the calculation of the survival function when $\alpha = \beta \exp(-\theta t)$, $\theta = 0.027$, and $\beta = 5, 15, 60$. To get the value of θ, note that for the no-sample case, the Bayes estimator is given by $\hat{\bar{F}}_0(t) = \exp(-\theta t)$. By making this estimator satisfy $\exp(-\theta T) = 0.543$, where T is the 0.543th quantile on the NPMLE curve, one obtains $\theta = 0.027$. Three values of β were considered to examine the effect of the

Table 6.3.1. Successive Minima, Plane 7912

i	X_i	K_i
1	23	3
2	7	11
3	5	4
4	3	8
5	1	4

Table 6.3.2. NPMLE and Bayes Estimator of $\overline{F}(t)$ for Data in Table 6.3.1

t in	NPMLE	Bayes Estimator, $\alpha(t) = \beta \exp(-\theta t)$				
(0, 1)	1.00	$\dfrac{\beta e^{-\theta}+30}{\beta+30}$				
[1, 3)	0.97	$\dfrac{\beta e^{-\theta}+26}{\beta+30}$	$\dfrac{\beta e^{-\theta}+29}{\beta+26}$			
[3, 5)	0.93	$\dfrac{\beta e^{-\theta}+18}{\beta+30}$	$\dfrac{\beta e^{-\theta}+29}{\beta+26}$	$\dfrac{\beta e^{-3\theta}+25}{\beta+18}$		
[5, 7)	0.88	$\dfrac{\beta e^{-\theta}+14}{\beta+30}$	$\dfrac{\beta e^{-\theta}+29}{\beta+26}$	$\dfrac{\beta e^{-3\theta}+25}{\beta+18}$	$\dfrac{\beta e^{-5\theta}+17}{\beta+14}$	
[7, 23)	0.82	$\dfrac{\beta e^{-\theta}+3}{\beta+30}$	$\dfrac{\beta e^{-\theta}+29}{\beta+26}$	$\dfrac{\beta e^{-3\theta}+25}{\beta+18}$	$\dfrac{\beta e^{-5\theta}+17}{\beta+14}$	$\dfrac{\beta e^{-7\theta}+13}{\beta+3}$
[23,∞)	0.55	$\dfrac{\beta e^{-\theta}}{\beta+30}$	$\dfrac{\beta e^{-\theta}+29}{\beta+26}$	$\dfrac{\beta e^{-3\theta}+25}{\beta+18}$	$\dfrac{\beta e^{-5\theta}+17}{\beta+14}$	$\dfrac{\beta e^{-7\theta}+13}{\beta+3}$ $\dfrac{\beta e^{-23\theta}+13}{\beta+3}$

size of β on the Bayes estimator. It was observed that the larger the value of β, the smoother the Bayes estimate compared to the NPMLE. In the extreme case when $\beta \to \infty$, the estimator reduces to $\exp(-\theta t)$ with no effect of the data. Table 6.3.2 displays the Bayes estimator.

As with the other nonparametric models, Tiwari and Zalkikar also extend the single sample case to multiple independent samples. As before, assume that record-breaking samples are obtained from m i.i.d. samples each of size n. The combined ordered record-breaking data are once again given by $\{x_{(i)}, i = 1, 2, \ldots, r^*\}$, and $\{k_{(i)}, i = 1, 2, \ldots, r^*\}$ denote the k_{ij}s corresponding to the $x_{(i)}$s. To calculate the Bayes estimator, the data are again redefined as $\{(Z_{(i)}, \delta_{(i)}), 1 \le i \le mn\}$, where $Z_{(i)}$s are the ordered Z_{ij} values in the m samples combined (the Z_{ij}s are defined as for the single sample case) and the $\delta_{(i)}$s are the corresponding induced values on the δ_{ij}s. Then the Bayes estimator of the survival function $\bar{F}(t)$ (based on squared-error loss and the Dirichlet process prior) is given by

$$\hat{\bar{F}}_{\alpha,m}(t) = B_{n,\,m}(t)\, W_{n,\,m}(t),\ t > 0, \tag{6.3.2}$$

where

$$B_{n,\,m}(t) = \frac{\alpha(t) + N_{n,m}^{+}(t)}{\alpha(R^{+}) + nm} \tag{6.3.3}$$

and

$$W_{n,\,m}(t) = \prod_{j=1}^{nm}\left(\frac{\alpha(Z_{(j)}) + N_{n,m}^{+}(Z_{(j)}) + \lambda_{(j)}}{\alpha(Z_{(j)}) + N_{n,m}^{+}(Z_{(j)})}\right)^{I\left(Z_{(j)} \le t,\ \delta_{(j)} = 0\right)/\lambda_{(j)}}. \tag{6.3.4}$$

As in the single sample case, $\alpha(u) = \alpha(u,\ \infty)$, $N_{n,m}^{+}(t)$ is the total number of $Z_{(j)}$s greater than t, and $\lambda_{(j)}$ is the multiplicity of $Z_{(j)}$.

Tiwari and Zalkikar have established weak convergence for the estimator $\hat{\bar{F}}_{\alpha,m}(t)$ in (6.3.2) as shown in the next theorem.

Theorem 6.3.2: (Asymptotic Normality of $\hat{\bar{F}}_{\alpha,m}(t)$) (Tiwari and Zalkikar, 1991) For all T such that $F(T) < 1$, the process $\{m^{1/2}(\hat{\bar{F}}_{\alpha,m}(t) - \bar{F}(t)): 0 \le t \le T\}$ converges weakly to a mean zero Gaussian process with covariance function

$$C(s,t) = \bar{F}(s)\bar{F}(t)\int_{0}^{s}\frac{1}{n\,H(x)}\,\frac{1}{\bar{F}(x)}\,dF(x),\ 0 < s \le t \le T,$$

where

$$H(t) = \frac{1}{n}\sum_{j=1}^{n}(1 - F(t))^{j}.$$

Proof: As in Samaniego and Whitaker (1988), the proof is based on showing the asymptotic equivalence of $\hat{\Lambda}_{\alpha,m} = -\ln\hat{\bar{F}}_{\alpha,m}(t)$ and Λ^{*}, which is defined by

$$\Lambda^* = \int \frac{1}{H_m} \, d(H_m + \tilde{H}_m),$$

where

$$H_m(t) = \frac{1}{mn} \sum_{i=1}^{m} \sum_{j=1}^{n} I[Z_{ij} > t]$$

and

$$\tilde{H}_m(t) = \frac{1}{mn} \sum_{i=1}^{m} \sum_{j=1}^{n} I[Z_{ij} \le t, \ \delta_{ij} = 0].$$

Now, from (6.3.3), we have

$$|B_{n,m}(t) - H(t)| = \left| \frac{\alpha(t) - \alpha(R^+)H_m(t)}{\alpha(R^+) + mn} \right| \le \left| \frac{\alpha(R^+)(n - H_m(t))}{n(\alpha(R^+) + mn)} \right|.$$

Since $nH_m(t)$ is the average of m i.i.d random variables, $I[Z_{ij} > t]$, by invoking the strong law of large numbers, we have that

$$\sqrt{m} \ \sup_{0 \le t \le T} |B_{n,m}(t) - H(t)| \to 0$$

with probability one as $m \to \infty$. Hence,

$$\sqrt{m} \ \sup_{0 \le t \le T} |\ln B_{n,m}(t) - \ln H(t)| \to 0$$

with probability one as $m \to \infty$.

Next, turning our attention to $W_{n,m}(t)$ given by (6.3.4), we have

$$\left| -\ln W_{n,m}(t) + \int_0^t \frac{1}{H_m(x)} \, d\tilde{H}_m(x) \right| \le |A_{m,1}(t)| + |A_{m,2}(t)|,$$

where

$$|A_{m,1}(t)| = \left| -m \int_0^t \frac{1}{\alpha(x) + mn H_m(x)} \, d(n\tilde{H}_m(x)) + \int_0^t \frac{1}{H_m(x)} \, d(n\tilde{H}_m(x)) \right|$$

$$\leq \frac{\alpha(R^+)\, \tilde{H}_m(T)}{H_m(T)(\alpha(T) + mn\, H_m(T))}$$

and

$$A_{m,2}(t) \leq \left(\frac{mn^2}{(\alpha(T) + mn\, H_m(T))^2}\right)\left(\frac{\tilde{H}_m}{(\alpha(T) + mn\, H_m(T) - 1)}\right).$$

Again, since $n\, \tilde{H}_m(t)$ is the sum of i.i.d random variables, applying the strong law of large numbers to $H_m(t)$ and $\tilde{H}_m(t)$, we have that

$$\sqrt{m}\ \sup_{0 \leq t \leq T} |A_{m,k}(t)| \to 0$$

with probability one as $m \to \infty$ for $k = 1, 2$. Thus,

$$\sqrt{m}\ \sup_{0 \leq t \leq T} |\hat{\Lambda}_{\alpha, m} - \Lambda^*| \to 0$$

with probability one as $m \to \infty$. This result, combined with Lemma 3.2 and Theorem 4.1 of Samaniego and Whitaker (1988), completes the proof of the theorem. ◆

Another problem of interest is in the empirical Bayes setup (which is related to the process of observing repeated samples of records), which is to estimate the survival curve at the $(m + 1)$st stage, based on m i.i.d. samples of record-breaking data. For simplicity, Tiwari and Zalkikar assume that the observed record values at all of the $(m + 1)$ stages are given by the sequence $\{(X_{il}, K_{il}), i = 1, 2, \ldots, m + 1\}$. The Bayes estimator of the survival function at the $(m + 1)$st stage under the squared-error loss function is then given by

$$(M + n)\, \hat{\bar{F}}_\alpha^{m+1}(t) = M\,\bar{\alpha}(t) + n\, I[X_{m+1,1} > t] + \frac{(n-1)\bar{\alpha}(t)}{\bar{\alpha}(X_{m+1,1})}\, I[X_{m+1,1} \leq t],$$

$$(6.3.5)$$

where $\bar{\alpha}(t) = \alpha(t)/M$ and $M = \alpha(R^+)$.

When M is known, Tiwari and Zalkikar propose replacing $\bar{\alpha}(t)$ in (6.3.5) by its estimator $\hat{\alpha}(t)$ to get the empirical Bayes estimator

$$(M + n)\, \hat{\bar{F}}_\alpha^{m+1}(t) = M\,\hat{\alpha}(t) + n\, I[X_{m+1,1} > t] + \frac{(n-1)\hat{\alpha}(t)}{\hat{\alpha}(X_{m+1,1})}\, I[X_{m+1,1} \leq t],$$

$$(6.3.6)$$

where

$$\hat{a}(t) = \frac{N(t)}{m} \prod_{i=1}^{m} \left(\frac{N(x_{i1}) + 2}{N(X_{i1}) + 1} \right)^{I(X_{i1} \le t)} \tag{6.3.7}$$

with $N(t)$ being the number of X_{i1}s greater than t.

The sequence of empirical Bayes estimators $\{ \hat{\bar{F}}_\alpha^{m+1} \}$, $m = 1, 2, \ldots$ defined in (6.3.6) (for M known) has been shown to be asymptotically optimal with a rate of convergence of $O(m^{-1})$.

Example 6.3.2. (Tiwari and Zalkikar, 1991)
To illustrate the calculation of the empirical Bayes estimator at the (M + 1)st stage, consider a variation of the Boeing 720 airplane air conditioner failure data of Proschan (1963). To easily follow the calculations, Tiwari and Zalkikar (1991) consider only the first pair (x_{i1}, k_{i1}), $i = 1, 2, \ldots, 10$, of the record data corresponding to each of the 10 planes (7908-7915, 8044, 8045). The data areas follows.

X_{i1}:	413	90	74	55	23	97	50	359	487	102
K_{i1}:	1	1	1	6	3	1	1	1	1	2

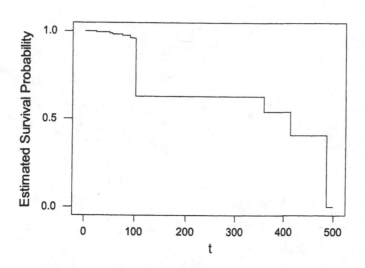

FIGURE 6.1. Empirical Bayes Estimate of Air Conditioner Survival Function with m = 9.

Note that the pairs (X_{i1}, K_{i1}), i = 1, 2, . . . , 10, are independent. For the empirical Bayes estimator, they assume that $n = 2$, $M = 1$, and treat the last pair of observations as the ones obtained at the $(m + 1)$st stage ($m =$ 9). Using equation (6.3.6) with $\hat{\alpha}$ as defined in (6.3.7), they obtain the survival function estimate $\hat{\bar{F}}_\alpha^{m+1}$ as a step function with jumps at the following points (the jump sizes are given in parentheses):

23(0.004), 50(0.006), 55(0.006), 74(0.009), 90(0.012), 97(0.002), 102(0.333), 359(0.086), 413(0.135), 487(0.407).

The estimated survival function is shown in Figure 6.1.

7
Record Models with Trend

7.1 Introduction

Weather patterns are constantly changing, athletic performances improve over time, and crime rates in cities increase or decrease over the years—in other words, populations change, and sometimes quickly! These situations, however, are typical of the ones for which we want to predict the next record. As examples, consider the following. Florence Griffith Joyner holds the record for the fastest speed for the women's 100-meter dash at the Olympics, clocking in at 10.62 seconds in 1988. Who is going to break her record and by how much? Inge De Bruijn from the Netherlands set the 100-meter record for both the backstroke and the butterfly at the 2000 Summer Olympics. Is he going to break his own record at the next Olympics? Is he going to clock in at less than 50 seconds for the backstroke the next time?

Unfortunately, we know that records drawn from changing populations cannot be explained by the *classical record model*, which requires observing records from an independent identically distributed sample. As a result, a number of statisticians have begun studying record models that allow for a changing population or trend. Three main types of models are identified in this context by Arnold et al. (1998): (i) a model that incorporates a geometrically increasing population (a more general case of this model is called the F^α model), (ii) a model that incorporates a linear trend in the underlying population (called the linear trend model), and finally, (iii) the Pfeifer model, so named since it was developed by Pfeifer (1982), who suggested allowing for an underlying population that changes after every record event.

We define the various models mentioned above in Section 7.2. The geometric model and the Pfeifer model have been presented in great detail by Arnold et al. (1998). As a result, these models are not discussed in much detail in this chapter. A brief review of some of the properties and applications of these models, however, is presented in Section 7.3. The linear drift model, especially the versions not covered by Arnold et al. (1998), are presented in greater detail in Section 7.4. In Section 7.5, under the heading of parametric *general record models*, we review the techniques of Carlin and Gelfand (1993), providing a completely general approach to predicting future re-cords. The only requirement of this approach is the specification of the joint distribution of the entire data sequence and the index set of the record-breaking observations. With the joint distribution completely specified, parameters are estimated using Monte Carlo techniques, and then the fitted model is used for prediction.

7.2 The Models for Records with Trend

The study of record models incorporating trend was first undertaken by Foster and Stuart (1954) who used the resulting record sequence to test for trend in location or in variance. However, the use of models with trend to predict future records was first considered by Yang (1975) who studied records from a population that grows at a geometric rate, thus giving rise to the "geometric model." The principles of Yang's model as well as more general cases thereof (and called F^α models) were studied extensively by Nevzorov in a series of papers (see Nevzorov, 1987, for example). The F^α models were then followed by the "linear drift" models of Ballerini and Resnick (1985, 1987) and Smith (1988), among others. In most of the linear models, the authors adopted the same framework as the classical model, retaining the independence assumption, but allowing for a trend in location. The models in most of the papers mentioned above were parametric in nature. The underlying population and the trend were completely specified parametrically, and then parameters were estimated by standard techniques. As mentioned in the introduction, these specific parametric models were then followed by the general parametric model of Carlin and Gelfand (1993). There have also been some recent advances made in the linear trend model. Feuerverger and Hall (1996) proposed nonparametric approaches to the linear trend model, and Borokov (1999) developed characterization-type results for the linear model and also considered Markov chains related to the process of obtaining records.

Some of the above-mentioned results are presented in the following sections. We start with the definitions of the models.

7.2.1. The F^α Model

Let $\{Y_n, n \geq 1\}$ be a sequence of independent continuous random variables, where the c.d.f of Y_n has the form $F_n(y) = \{F(y)\}^{\alpha(n)}$, $\alpha(n) > 0$, for some c.d.f. F. Then the corresponding record value sequence based on the Y_ns, X_1, K_1, X_2, K_2, $\ldots$, is referred to as the "F^α record model." Note that when $\alpha(n) = 1$ for all n, we have the *classical record model*. Yang (1975) considered the F^α record model where the underlying population was geometrically increasing and showed that for this model, the asymptotic distribution of the interrecord times was also geometric. Yang used the geometric model to fit Olympic records, albeit, not very well. As mentioned earlier, Nevzorov studied the properties of the F^α model (in fact, he coined the name for the models) in a series of papers and showed that although this model includes a number of situations encountered under the classical model, the asymptotic properties of this model depend on the sequence $\alpha(n)$. The results are described in more detail in Section 7.3.

7.2.2 The Pfeifer Model

In the F^α model, one assumes that each successive observation comes from a different population. The Pfeifer model provides a slightly different version of this approach. Here, the population is assumed to change after the occurrence of each record event rather than each observation, described as follows by Pfeifer (1982).

Assume that $\{Y_{00}, Y_{nk}; \; n, \; k \geq 1\}$ is a sequence of independent random variables on some probability space $(\Omega, \mathcal{A}, P)$, with P_n being the probability distribution of the Y_{nk} and F_n being the corresponding distribution function. The sequence of interrecord times $\{K_n; \; n \geq 0\}$ is defined recursively by $K_0 = 0$, $K_{n+1} = \min\{k: Y_{n+1,k} > Y_{n,K_n}\}$. Clearly then, the sequence of record values $\{X_n: \; n \geq 0\}$ is defined by $X_n = Y_{n, K_n}$ and the sequence of the record times is given by $N_n = 1 + \sum_{j=0}^{n} K_j$.

Pfeifer (1982) uses the following example to illustrate the model. Suppose that the sequence $\{Y_{nk}; \; k \geq 1\}$ corresponds to random shocks attacking a component which works without failure unless a shock greater than X_{n-1} occurs. Suppose then that for safety reasons, a modified component is used which can endure shocks up to a magnitude of the last shock X_n. Let us also assume that safety factors are built in which influence the distribution of subsequent shocks $\{Y_{n+1,k}; \; k \geq 1\}$. Then N_n denotes the time to the nth failure, with the time between failures K_n stochastically increasing if the underlying distributions P_n are stochastically decreasing. This case corresponds to a shock model with increasing safety.

7.2.3. The Linear Drift Model

Consider the record value sequence, $X_1, K_1, X_2, K_2, \ldots$, obtained from a sequence of independent random variables $\{Y_n, n \geq 1\}$, such that $Y_n = Z_n + c_n$, where Z_n is a sequence of independent identically distributed random variables and c_n is a nonrandom trend. If c_n is linear in n, then the resulting model is called the *linear drift model*.

Arnold et al. (1998) show that if the distribution function of the Z_is is taken to be the Gumbel c.d.f., the resulting model is in fact an F^α model with $\alpha(n) = \exp(c_n)$, once again illustrating the usefulness of the model developed by Yang.

We now describe the properties of these models.

7.3   The Geometric and the Pfeifer Models

7.3.1   The Geometric Model

We start this section with a description of the development of Yang's geometric model and its properties. Recall that the classical record model assumes that the records are drawn from independent identically distributed observations, $Y_1, Y_2, \ldots$. In order to take into account the increasing underlying population in the case of Olympic games, Yang (1975) assumed that each of the Y_is themselves is the maximum of an increasing number n_i of independent and identically distributed random variables. In other words,

$$Y_i = \max (Z_{i1}, Z_{i2}, \ldots, Z_{in_i}),$$

where $\{Z_{ij}\}$, $1 \leq j \leq n_i$; $i = 1, 2, \ldots$, is a sequence of i.i.d. random variables. Then from Yang (1975), we have that for the case of the Olympic games, n_i would be the size of the world at the ith game, and Y_i would be the champion for that game. By further assuming that the population is increasing geometrically, that is, by letting $n_i = \lambda^{i-1} n_1$ for some $\lambda \geq 0$, we have the first version of the geometric model! (Note however, if λ is not an integer, then n_i has to be approximated by the integer closest to $\lambda^{i-1} n_1$; that is, we let $n_i = \lfloor \lambda^{i-1} n_1 + 0.5 \rfloor$, where $\lfloor X \rfloor$ denotes the greatest integer less than or equal to X.) Let X_i, K_i, and N_i denote the sequence of record values, interrecord times, and the record times, respectively, obtained from the above geometric model. Yang (1975) has established the following results for the interrecord time sequence K_i, the first of which we state without proof.

Theorem 7.3.1.1: Define the series $S(k)$ as the total size of the population associated with the sample $Y_1, Y_2, \ldots, Y_k$. That is,

$$S(k) = n_1 + n_2 + \cdots + n_k.$$

Then regardless of the form of the n_is, we have that

$$P(K_1 > j) = n_1/S(j + 1)$$

and

$$P(K_n > j) = \sum_{w_1=2}^{\infty} \sum_{w_2=w_1+1}^{\infty} \cdots \sum_{w_{n-1}=w_{n-2}+1}^{\infty} \frac{n_1 n_{w_1} \ldots n_{w_{n-1}}}{S(w_1 - 1) \ldots S(w_{n-1} - 1) S(w_{n-1} + j)}.$$

Although the details of the proof are cumbersome in notation, the theorem is easily proved by using the fact that the distribution of the interrecord times is distribution free, since it depends only on the ranks of the Y_is. Thus without loss of generality, one can assume that the X_{ij}s have the exponential distribution and then prove the theorem using well-known results on the distribution of the record values and the record times.

To use the above theorem for the geometric model, define the following.

$$\sigma(j) = 1 + \lambda + \cdots + \lambda^{j-1}$$

$$p(n, 0) = 1, \quad p(1, j) = 1/\sigma(j + 1),$$

and

$$p(n, j) = \sum_{w_1=2}^{\infty} \sum_{w_2=w_1+1}^{\infty} \cdots \sum_{w_{n-1}=w_{n-2}+1}^{\infty} \left\{ \prod_{i=1}^{n-1} \frac{\lambda^{w_i-1}}{\sigma(w_i - 1)} \right\} \frac{1}{\sigma(w_{n-1} + j)},$$

$$\text{for } j \geq 1. \qquad (7.3.1.1)$$

Then from Theorem 7.3.1.1, we have the following lemma.

Lemma 7.3.1.1: Let $p(n, j)$ be defined as in (7.3.1.1) and let $q_j = \lim_{j \to \infty} p(n, j)$. Then

$$p(n, j) = \sum_{i=0}^{j} p(n-1, i)/\sigma(j + 1) \qquad (7.3.1.2)$$

and q_j exists and is given by

$$q_j = \lambda^{-j}, \quad j \geq 0.$$

Proof: The proof of the first part of Lemma 7.3.1.1 follows easily from replacing the last summation in equation (7.3.1.1) by the equality

$$\sum_{w_{n-1}=w_{n-2}+1}^{\infty} \frac{\lambda^{w_{n-1}-1}}{\sigma(w_{n-1}-1)} \frac{1}{\sigma(w_{n-1}+j)} = \frac{1}{\sigma(j+1)} \sum_{i=0}^{j} \frac{1}{\sigma(w_{n-1}+i)}.$$

For the second part of the lemma, note that from (7.3.1.2), it follows easily that $p(2, j) \geq p(1, j)$, for all j. It also can be shown that

$$P(K_{n+1} > j) - P(K_n > j) = \sum_{i=1}^{j} \{P(K_n > i) - P(K_{n-1} > i)\} / \sigma(j+1),$$

thereby establishing the monotonicity of $p(n\ j)$, which in turn then guarantees the existence of a limit for $p(n, j)$. The computation of q_j follows easily from (7.3.1.2). ♦

From Theorem 7.3.1.1 and Lemma 7.3.1.1, one can also easily see that for the geometric model $p(n, j) = P(K_n > j)$ if λ is an integer. For a noninteger λ, Yang (1975) has the following bound on the probability (stated in the form of a lemma without proof).

Lemma 7.3.1.2: For any $\lambda > 1, j \geq 1$, we have

$$|P(K_n > j) - p(n,j)| \leq \frac{c}{n_1} \sum_{i=1}^{n} \{2/(1+\lambda)\}^{n-i} \lambda^{-i}, \tag{7.3.1.3}$$

where $c = 32\lambda^2 / (\lambda^2 - 1)$.

The above lemma then gives the main result for Yang's model.

Theorem 7.3.1.2: For the geometric record model, let

$$p_j = \lim_{n \to \infty} P(K_n = j).$$

Then

$$p_j = (\lambda - 1)\lambda^{-j}, \quad j = 1, 2, \ldots;$$

that is, the asymptotic distribution of the interrecord times for a geometric model is geometric with $p = (\lambda - 1)/\lambda$.

Proof: If $\lambda = 1$, then we have the classical record model and the theorem follows from the results for the classical model. For $\lambda > 1$, note that the right-hand side of (7.3.1.3) tends to zero and n goes to infinity. Thus we have

$$\lim_{n \to \infty} P(K_n = j) = \lim_{n \to \infty} (P(K_n > j - 1) - P(K_n > j))$$

$$= q_{j-1} - q_j = (\lambda - 1) \lambda^{-j}$$

from Lemma 7.3.1.1. ♦

These results led Yang (1975) to conjecture that if one had a geometric population and one knew its growth rate, one could then estimate the interrecord times for such a population. As an application, Yang (1975) used the geometric model to fit the breaking times between Olympic records.

As the first step in estimating interrecord times for a geometric population, one needs to estimate λ. Noting that there have been Olympic games every four years since 1896 (except in the war years), that the world's population has approximately doubled every 36 years, and one has 9 four-year periods in 36 years, we have $\lambda^9 = 2$; that is, $\lambda = 1.08$.

Hence the limiting distribution of the interrecord times for the Olympic games is geometric with $p = 0.08/1.08 = 0.074$. This indicates that we would expect the waiting time between successive records to be approximately $1/0.074$ or 13.5 games. But, in fact, records are being broken much faster than every 13.5 games. Hence Yang (1975) ends the paper by concluding that a geometric growth in the population alone cannot explain the rapid breaking of Olympic records.

The geometric model of Yang (1975) evolved into a more general model under Nevzorov, who published a series of papers on it in the 1980s. That model incorporates a variety of situations for which the underlying sequence consists of non-identically distributed random variables. However, at the same time, it also retains the independence assumption of the classical record model. The properties of this model, both finite-sample and asymptotic, have been widely studied by Nevzorov (see, for example, Nevzorov 1985, 1990, 1995), as well as by Ballerini and Resnick (1987) and Arnold et al. (1998). The use of the model to explain real sets of data, however, has been limited.

7.3.2. The Pfeifer Model

The model developed by Pfeifer in 1982 has already been described in Section 7.2.2. Let the sequence $\{Y_{00}, Y_{nk}; n, k \geq 1\}$, the probability space $(\Omega, \mathcal{A}, P)$, P_n, and F_n be defined as before. Recall then that the sequence of interrecord times, the record values, and the record times were denoted as usual by $\{K_n; n \geq 0\}$, $\{X_n; n \geq 0\}$ and $\{N_n; n \geq 0\}$, respectively. Pfeifer (1982) considered only the nondegenerate record model, that is, the model for which the interrecord time $K_n < \infty$, and shows that for this model, the upper records form a Markov chain.

In this subsection, we state some of the properties of the Pfeifer model in the next two theorems. These theorems are stated without proof since in most cases, the proof is similar to the classical case.

Theorem 7.3.2.1: The set $\{(K_n, X_n); n \geq 0\}$ is a Markov chain with transition probabilities

$$P_{n-1, n}(m, x \mid A \times B) = P_n((x, \infty) \cap B) \sum_{j \in A} F_n^{j-1}(x), \qquad (7.3.2.1)$$

$m \in N$, $x < \xi_n$ (here, ξ_n is the right endpoint of the distribution function F_n), $A \subseteq N$, and $B \in \mathfrak{B}$, where $\mathfrak{B}$ is the collection of all Borel sets $B \subseteq R$ and N denotes the nonnegative integers.

From the above, it also follows that the set $\{(N_n, X_n); n \geq 0\}$ is also a Markov chain with transition probabilities

$$Q_{n-1,n}(m, x \mid A \times B) = P_n((x, \infty) \cap B) \sum_{j \in A \cap (m,\infty)} F_n^{j-m-1}(x), \qquad (7.3.2.2)$$

$m \in N$, $x < \xi_n$, $A \subseteq N$, and $B \in \mathfrak{B}$, which are translation invariant with respect to m.

The main result of this section is then an immediate consequence of (7.3.2.1) and (7.3.2.2).

Theorem 7.3.2.2:
a) The set $\{(N_n, X_n); n \geq 0\}$ is a Markov additive chain and the sequence $\{X_n: n \geq 0\}$ is a Markov chain with transition probabilities

$$P_{n-1,n}(x \mid B) = P_n(B \mid (x, \infty)), \quad x < \xi_n, B \in \mathfrak{B}.$$

b) $K_1, K_2, \ldots, K_n$ are conditionally independent given $X_0, X_1, \ldots, X_{n-1}$ with

$$P\left(\bigcap_{i=1}^{n}\{K_i = m_i\} \mid X_0, \ldots, X_{n-1}\right) = \prod_{i=1}^{n} P(K_i = m_i \mid X_{i-1})$$

$$= \prod_{i=1}^{n} \{1 - F_i(X_{i-1})\} F_i^{m_{i-1}}(X_{i-1})$$

a.s. $m_1, m_2, \ldots, m_n \in N.$

7.4 Properties of the Linear Drift and Related Models

7.4.1 Early Work

As mentioned in Section 7.2, the record value sequence, X_1, K_1, X_2, K_2, . . . , is called a *linear drift record model* if it is obtained from a sequence of independent random variables $\{Y_n, n \geq 1\}$ such that

$$Y_n = Z_n + c_n, \tag{7.4.1.1}$$

where Z_n is a sequence of i.i.d. random variables and c_n is a linear function in n. Linear drift and other related models were developed after the F^α models. As mentioned earlier, Yang (1975) used a geometrically increasing population to try to explain Olympic records, but then concluded that such a model underestimated the rate at which Olympic records were being broken. As a result, Ballerini and Resnick (1985) proposed a record model based on improving populations. We consider their model briefly here.

In (7.4.1.1), let the sequence $c_n = cn$, where the constant c is greater than 0 if upper records are of interest and less than 0 if one is instead interested in lower records. Ballerini and Resnick (1985) studied some properties of records obtained from (7.4.1.1) both when the sequence $\{Z_i\}$ consists of i.i.d. random variables and also when the sequence $\{Z_i\}$ is strictly stationary. In particular, they showed that the record rate for a "linear trend" model is almost surely asymptotically linear when the sequence $\{Z_i\}$ is strictly stationary. In addition, when the sequence $\{Z_i\}$ is i.i.d., then the record rate is asymptotically normally distributed as well. The next two theorems of Ballerini and Resnick give the properties of the N_n, the number of records in a sequence of size n, and the corresponding record rate N_n/n.

Theorem 7.4.1.1: Suppose that the sequence $\{Z_i\}$ is i.i.d. with a common distribution function F, $c > 0$, and $EZ_1^+ < \infty$. Then as $n \to \infty$, $N_n/n \to p$ almost surely, where p is a constant between 0 and 1.

(*Note:* The result is the same for lower records provided $c < 0$, $EZ_1^- < \infty$, and F is replaced by $1 - F(-x)$. Moreover, the result of a constant asymptotic record rate also holds for a stationary sequence $\{Z_i\}$, although the limit, in general, will be different from the i.i.d. case.)

Proof: For $1 \leq j \leq n$, define the record indicator variables as

$$I_j = \begin{cases} 1, & \text{if } Y_j \text{ is a record} \\ 0, & \text{otherwise.} \end{cases}$$

First, we consider the rate for $j \geq 2$. Then

$$\begin{aligned}
E(I_j) &= P(Y_j \text{ is an upper record}) = P(Y_j > Y_i, \ 1 \leq i \leq j-1) \\
&= P\left(X_j + cj > X_i + c_i, 1 \leq i \leq j-1\right) \\
&= P\left[\bigvee_{k=1}^{j-1} (X_k - (j-k)c) < X_j\right] \\
&= \int_{-\infty}^{\infty} G_j(y)\, F(dy) \ \downarrow \ \int_{-\infty}^{\infty} G_\infty(y)\, F(dy) = p \ \text{ as } j \to \infty,
\end{aligned}$$

where

$$G_n(x) = \prod_{j=1}^{n-1} F(x + c_j),$$

$$G_\infty(x) = \prod_{j=1}^{\infty} F(x + c_j) = p,$$

the asymptotic rate of obtaining records, and "v" denotes "maximum."

Hence $\lim_{n \to \infty} E(I_n) \downarrow p$ which then implies that $n^{-1}EN_n \to p$. The next step is to prove L_2-convergence.

Note that

$$E(n^{-1}N_n - p)^2 = E(n^{-1}N_n)^2 - 2pn^{-1}EN_n + p^2$$

$$= E(n^{-1}N_n)^2 - p^2 + o(1).$$

Thus we have to show that $E(n^{-1}N_n)^2 \to p^2$. First,

$$E(n^{-1}N_n)^2 = \frac{1}{n^2} \sum_{i=1}^{n} \sum_{j=1}^{n} EI_i I_j = o(1) + 2n^{-2} \sum_{i=1}^{n} \sum_{j=1, i \leq j}^{n} EI_i I_j.$$

Then

$$\sum_{i=1}^{n} \sum_{m=1}^{n} EI_i I_{i+m} = P\left[Y_i \text{ is a record}, Y_{i+m} \text{ is a record}\right]$$

$$= P\left[\bigvee_{k=1}^{j-1} (X_k - (i-k)c) < X_i < X_{i+m} + cm, \bigvee_{k=1}^{m-1} (X_{i+k} - (m-k)c) < X_{i+m}\right]$$

$$= \iint_{y < s + cm} G_i(y)\, G_m(s)\, F(ds)\, F(dy)$$

$$= \int_{-\infty}^{\infty} F(dy)\, G_i(y) \int_{s=y-cm}^{\infty} F(ds)\, G_m(s).$$

Thus

$$\lim_{m \to \infty} EI_i I_{i+m} = p \int_{-\infty}^{\infty} F(dy)\, G_i(y)$$

and so

$$\lim_{i \to \infty} \lim_{m \to \infty} EI_i I_{i+m} = p \int_{-\infty}^{\infty} F(dy)\, G_i(y) = p^2.$$

Therefore we have the L_2-convergence that was needed.

The almost sure convergence is shown by using the subadditive ergodic theory by Kingman (1973). For nonnegative integers $0 \leq s < t$, define

$$\xi_{s,t} = 1 + \sum_{j=s+2}^{t} I_{[Y_j \,>\, \max\,(Y_{s+1},\, \ldots,\, Y_{j-1})]}\,;$$

that is, $\xi_{s,t}$ is the number of records in the segment $\{Y_{s+n},\ 1 \leq n \leq t - s\}$. Note that $\xi_{0,n} = N_n$. Now $\lim_{n \to \infty} n^{-1}\xi_{0,n}$ exists almost surely, provided Kingman's three subadditive axioms are satisfied. These axioms are as follows.

S1: For $0 \leq s < u < t$, we require that $\xi_{s,t} \leq \xi_{s,u} + \xi_{u,t}$. But this is obvious, since a record among $Y_{s+1}, \ldots, Y_t$ will be either a record among $Y_{s+1}, \ldots, Y_u$ or among $Y_{u+1}, \ldots, Y_t$.

S2: We need to show that $\xi_{s,t}$ is equal in distribution to $\xi_{s+1,t+1}$. We have

$$\xi_{s,t} = 1 + \sum_{j=s+2}^{t} I_{[Y_j + cj > \max(Y_i + ci,\, s+1 \leq i \leq j-1)]}$$

$$\overset{d}{=} 1 + \sum_{j=s+2}^{t} I_{[Y_{j+1} + cj > \max(Y_i + ci,\, s+1 \leq i \leq j-1)]}$$

$$= 1 + \sum_{j=s+2}^{t} I_{[Y_{j+1} + c(j+1) > \max(Y_i + c(i+1),\, s+1 \leq i \leq j-1)]}$$

$$= 1 + \sum_{j=(s+1)+1}^{t+1} I_{[Y_j + cj > \max(Y_i + c(i+1), s+1 \le i \le j-1)]}$$

$$= \xi_{s+1,t+1} .$$

S3: Lastly, we need to show that $E \xi_{0,t} \ge -At$; but this follows automatically from the fact that $E \xi_{0,t} \ge 0$. The proof is now complete. ◆

The next theorem is stated without proof and provides the conditions for the asymptotic normality of the record rate.

Theorem 7.4.1.2: *(Asymptotic Normality of N_n)* In addition to the conditions of Theorem 7.4.1.1, assume that $E(Z_1^2) < \infty$. Then $\sqrt{n} (N_n / n - p)$ converges weakly to a mean zero normal random variable with variance $\sigma^2 = p - p^2 + 2 \sum_{m=1}^{\infty} (r_m - p^2)$, where

$$r_m = P\left(Y_i > \bigvee_{l<i} Y_l, Y_{i+m} > \bigvee_{l<i+m} Y_l \right)$$

for any $-\infty < i < \infty$ and p as defined earlier.

Assuming that the sequence $\{Z_i\}$ is strictly stationary, Ballerini and Resnick (1985) successfully modeled record times in mile-run data using (7.4.1.1). The fastest time in the mile was recorded for each year from 1860 until 1982 by Mengoni (1973). Ballerini and Resnick (1985) used a linear model with ARMA(1,1) residuals to model these times. Realizations simulated from the estimated model gave sample paths that were similar to the original model, illustrating a good fit.

The work of Ballerini and Resnick (1985) was followed by that of Smith and Miller (1986). They developed a state space model for athletic records, treating the observed data as censored observations. Prediction was done within a Bayesian framework, and results were applied to some athletic data. We consider this model very briefly here.

The original formulation of a lower record model assumes that we have a random sample of observations $Y_1, Y_2, \ldots, Y_N$, but we do not observe the whole sequence, observing instead only successive minimum values. When applying this to athletic records, say, running, for example, Smith and Miller (1986) argue that the random variable Y_i denotes the best performance in year i and then the records are the successive minima of $\{Y_i\}$. Statistical modeling for the series then, as proposed by Smith and Miller (1986), assumes that for the year in which a record is not broken, the best performance for that year is a censored observation. We know only that it is bigger (or smaller) than the record value.

A general state space model assumes that one has an underlying parameter sequence θ_n that evolves stochastically and that the observed sample $\{X_n, 1 \leq n \leq N\}$ has a density that depends on θ_n. However, the general model is fraught with difficulties in that the exact formulation of the distributions of θ_n and the X_ns is often intractable. To avoid these difficulties, the state space model used by Smith and Miller (1986) specifically assumes the following.

1. The underlying parameter θ_0 has a prior gamma distribution with parameters α and β.

2. For a fixed n, $1 \leq n \leq N$, conditional on θ_{n-1} and $X^{n-1} = \{X_i: 1 \leq n \leq n - 1\}$, $\theta_n = \theta_{n-1}\rho_n\xi_n$ where ρ_n is a nonrandom quantity, ξ_n has a beta distribution with shape parameter $\gamma_n\alpha_{n-1}$ and scale parameter $(1 - \gamma_n)\times$ α_{n-1} for some nonrandom quantity γ_n. Both ρ_n and γ_n depend on the X^{n-1} although not through θ_{n-1}.

3. Conditional on θ_n, X_n has an exponential distribution with parameter θ_n.

The above assumptions allow the development of recursive updating relations for the state space model. However, to generalize the model further, Smith and Miller (1986) assume that what one observes is not the X_n, but Y_n, where Y_n is related to X_n by a one-to-one transformation, (say, $X_n = T(Y_n \mid \phi_n)$ depending on yet another parameter ϕ_n). The parameters ϕ_n, ρ_n, and γ_n were estimated through Bayesian methods or through maximum likelihood techniques, and a predictive density was developed for future observations given past data.

Smith and Miller (1986) then argue that since the Y_is themselves represent a sequence of maxima or minima, one can assume an underlying Type I extreme value distribution for the Y_is. The series Y_1, $Y_2, \ldots, Y_N$ is then transformed to $X_1, X_2, \ldots, X_N$ by the transformation $X_n = \exp(\pm Y_n / \phi)$ (the $+$ sign taken with minima and the $-$ sign taken with maxima). Assuming further that $\gamma_n \equiv 1$ for all n and $\rho_n \equiv \rho$, the predictive distribution for the future record Z_{n+m} (given past n records) as developed by Smith and Miller (1986) is

$$P\{Z_{n+m} > y \mid X_N, \phi, \rho\} = \left[1 + \beta_N^{-1} \exp(y/\phi)\rho(\rho^m - 1)/(\rho - 1)\right]^{-\alpha N}.$$

The predictive distribution of a future record was also developed without the restriction that $\gamma = 1$, and the resulting model was used to fit mile and marathon records for male athletes. Tables 7.4.1 and 7.4.2 give the records and the model predictions as developed by Smith and Miller (1986).

Smith and Miller (1986) (as can be seen in Table 7.4.2) found that although the predictions looked reasonable for mile records, they were definitely too low for marathon records. Hence, as noted by Smith (1988), the work mentioned above allowed for some kind of dependence structure within the general record framework, however, it was restrictive in that it was based on just one underlying distribution which clearly did not explain the records very well. This led to the development of a class of trend models using different underlying populations and trends by Smith (1988). We next discuss some of that work.

Table 7.4.1. Men's World Mile and Marathon Records

Mile Records (record time in seconds)					
Year	Record	Year	Record	Year	Record
1931	249.2	1954	237.9	1975	229.4
1933	247.6	1957	237.2	1979	229.0
1934	246.7	1958	234.5	1980	228.8
1937	246.4	1962	234.4	1981	227.3
1942	244.6	1964	234.1	1985	226.3
1943	242.6	1965	233.5		
1944	241.6	1966	231.3		
1945	241.3	1967	231.1		
Marathon Records (record time in minutes)					
Year	Record	Year	Record	Year	Record
1909	162.52	1953	138.58	1969	128.57
1913	156.12	1954	137.67	1981	128.22
1920	152.6	1958	135.28	1984	128.08
1925	149.03	1963	134.47	1985	127.2
1935	146.70	1964	132.20		
1947	145.65	1965	132.00		
1952	140.72	1967	129.62		

Table 7.4.2. Predictions for the Mile and Marathon Records

	γ	ϕ	ρ	$\phi \ln(\gamma\rho)$	95%	50%	5%
				Percentiles of Predictive Distribution—Time Period 10 Years			
Mile	0.79	1.06	1.98	0.48	224.9	221.4	218.2
Marathon	0.84	1.98	1.64	0.640	124.1	118.9	113.1

7.4.2 Trend Models—The Work of Smith (1988) and Other Developments

Following the notation of Smith (1988), assume that the record value sequence $X_1, X_2, \ldots$, consists of *lower records* from the sequence

$$Y_n = Z_n + c_n, \tag{7.4.2.1}$$

where Z_n is a sequence of i.i.d. random variables and c_n is a nonrandom trend. The model is assumed to be parametric of the form $c_n = c_n(\beta)$ and $Z_n \sim f(x, \theta)$, where f is a continuous density function and β and θ are parameters. Maximum likelihood estimation of the vector (β, θ) is studied under various models.

The likelihood function of (β, θ) based on the lower records $X_1, X_2, \ldots, X_N$, is

$$\prod_{n=1}^{N} \{ f(X_n - c_n(\beta); \theta) \}^{\delta_n} \{ 1 - F(Z_n - c_n(\beta); \theta) \}^{1-\delta_n}, \tag{7.4.2.2}$$

where F is the corresponding c.d.f. and δ_n is 1 if the nth observation is a record and is 0 otherwise. The maximum likelihood estimates $(\hat{\beta}, \hat{\theta})$ are obtained by numerical methods.

These main distributions are considered for the underlying sequence Z_n:

(i) The normal distribution;

(ii) The Type I extreme value distribution (often referred to as the Gumbel distribution) with c.d.f. given by

$$F(x; \mu, \sigma) = 1 - \exp[-\exp\{(x - \mu)/\sigma\}], \; -\infty < \mu < \infty, \; \sigma > 0;$$

(iii) The Generalized Extreme Value Distribution (GEV) with c.d.f given by

$$F(x; \mu, \sigma, k) = 1 - \exp\left\{-\left(1 + k(x - \mu)/\sigma\right)^{1/k}\right\}, \quad 1 + \frac{k(x - \mu)}{\sigma} > 0$$

for $\mu \in \mathcal{R}$, and $\sigma > 0$.

For the trend, Smith considers these different functions:
a) A linear trend model: $c_n(\beta_0, \beta_1) = \beta_0 - n\beta_1$, $\beta_1 > 0$.
b) A quadratic-drift model: $c_n(\beta_0, \beta_1, \beta_2) = \beta_0 - n\beta_1 + n^2\beta_2/2$, $\beta_1 > 0$.
c) An exponential decay model: $c_n(\beta_0, \beta_1, \beta_2) = \beta_0 - \beta_1\{1 - (1 - \beta_2)^n\}/\beta_2$, $\beta_1 >$, $0 < \beta_2 < 1$.

Note that the combinations of the underlying distribution and the different types of trends give rise to nine different models under consideration: A: Normal, linear drift; B: Normal, quadratic drift; C: Normal, exponential decay; D: Gumbel, linear drift; E: Gumbel, quadratic drift; F: Gumbel, exponential decay; G: GEV, linear drift; H: GEV, quadratic drift; and I: GEV, exponential decay.

The models were fitted by numerical maximum likelihood to two data sets: the mile-run data considered by Smith and Miller (1986), and the marathon records data also tabulated by Smith and Miller (1986). A detailed analysis by Smith (1988) showed that the normal model was the most suitable of the three distributions. Moreover, since one expects the data to be linear over short periods of time, Smith (1988) fit Model A to different portions of the series. The results indicated that although the theory of a short-term linear trend was justified, over the whole series (1860 to 1930), a linear model does not fit the data very well. In fact, Smith found that a cubic regression fit the data better than any of the models considered earlier. To complete the analysis, Smith also fitted all nine models to the data. Moreover, since the initial plot had suggested a linear improvement from 1931 to 1985, Smith concentrated on this period in subsequent analysis. It turned out that in no case did the quadratic or the exponential decay model offer a significant improvement over the linear model. Comparing the distributions, Smith found that the normal distribution was clearly superior to the Gumbel, although the GEV distribution seemed to be the best of the three. However, the method of maximum likelihood used to fit the parameters of the GEV distribution did not always converge and so the GEV distribution was not pursued any further.

In conclusion then, Smith (1988) notes that over a short-term period, the linear drift model with normal errors fits athletic data well, but over long periods of time, the slope of the linear drift changes. Hence the use of such models for long-term forecasting is not recommended.

Smith's approach was further studied by Feuerverger and Hall (1996). They noted that the linear drift model is sensitive to misspecifications in the error distribution. In particular, they showed that the estimated parameters may not all be consistent if the error distribution is chosen incorrectly. As an example, if the errors are chosen to be normal when they in fact have some other distribution, then the estimator of the slope will be consistent, but the estimator of the intercept will not be consistent. Furthermore, the estimator of the slope will not converge at the rate predicted by the information matrix. Surprisingly though, this problem does not arise when the trend increases at a faster rate than the linear trend. In this case, the asymptotic properties of the estimators are unaffected by the choice of the error distribution.

To rectify this problem, Feuerverger and Hall (1996) proposed the use of nonparametric methods to estimate the slope and the intercept of the trend, the asymptotic variances of these estimates, and the error distribution. The estimators developed were "root n" consistent (n being the sample size). The estimator of the intercept is obtained via least squares fitting and has the same rate of convergence as the one obtained via maximum likelihood. The estimator of the slope has in fact a rate of convergence of $n^{-3/2}$. They used bootstrap methods to obtain the variance of the estimators and noted that in the case of linear trend, the variance of the estimator of the slope depends on the underlying density and may be estimated consistently, but not with the rate of $n^{-1/2}$.

Under the mantle of athletic records, a noteworthy article is one by Robinson and Tawn (1995) and the follow-up letter to the editor by Smith (1997). Robinson and Tawn (1995) developed a statistical model to assess whether an observed record is inconsistent with past data. Specifically, they analyzed the data over the period 1972 to 1992 in the women's 1500- and 300-meter running events with their goal being to examine more closely the performance of a Chinese athlete, Wang Junixa. She broke the world record by 6.08 seconds on September 13, 1993, having broken it the previous day by 10.43 seconds. It was suspected that she had taken performance-enhancing drugs, although she tested negatively for drug use. Hence the model was developed in the belief that statistical inconsistency of her performance with past data might be considered as evidence of use of performance-enhancing drugs. Robinson and Tawn (1995) fitted a model to the five best performances by athletes in each year and on the basis of their model showed that Wang's result was unlikely. Their paper also evoked a response from Smith (1997) who proposed a simpler Bayesian predictive method to analyze the same data and to show that Wang's results were indeed highly unlikely.

Finally, the stochastic properties of the "linear trend" model were also studied by Borokov (1999). Besides looking at the properties of the asymptotic rate function, Markov chains relating to the process of records were also considered and limit theorems were proven for them, including the "ergodicity" theorem. We state some of the main results of Borokov's paper in the form of two theorems.

To do so, let the sequence of interrecord times, the *upper record values*, and the record times be denoted by $\{K_n; n \geq 0\}$, $\{X_n; n \geq 0\}$, and $\{N_n; n \geq 0\}$. Further assume that the records are drawn from the model

$$Y_n = Z_n + cn,$$

where Z_n is a sequence of i.i.d. random variables with c.d.f. $F(z)$ and $c \geq 0$ is a nonrandom trend.

Theorem 7.4.2.1: (i) If $c = 0$, the sequences $\{X_n; n \geq 0\}$, $\{N_n; n \geq 0\}$, and $\{(K_n, X_n); n \geq 0\}$ form homogeneous Markov chains with transition probabilities given by the following relationships. For $n \geq 0$, $k > j \geq 1$, and $y \geq x \in (x_-, x_+)$,

$$P(K_{n+1} = k \mid K_n = j) = \frac{j}{k(k-1)}, \qquad P(X_{n+1} > y \mid X_n = x) = \frac{\overline{F}(y)}{\overline{F}(x)},$$

and

$$P(K_{n+1} = k, X_{n+1} > y \mid K_n = j, X_n = x) = \overline{F}(y) F^{k-j-1}(x).$$

(ii) If $c > 0$, then in the general case, the sequences $\{X_n; n \geq 0\}$ and $\{N_n; n \geq 0\}$ no longer form a Markov chain, but $\{(K_n, X_n); n \geq 0\}$ is still a homogeneous Markov chain with transition probabilities given by

$$P(K_{n+1} = k, X_{n+1} > y \mid K_n = j, X_n = x) = \overline{F}(y - ck) \prod_{j < m < k} F(x - cm).$$

(iii) If $E(Z_1^+) < \infty$ and $c > 0$, then $n^{-1}(K_n, X_n) \to (1, c)/p(c)$, where $p(c)$ is the asymptotic record rate.

For the next theorem, define the following,

$$\tau_n = \max\{K_m : K_m \leq n\},$$

the last time we had a record by time n, and

$$W_n = \left(W_n^{(1)}, W_n^{(2)}\right) \text{ with } W_n^{(1)} = n - \tau_n, \ W_n^{(2)} = Y_{\tau_n} - c\tau_n = Z_{\tau_n}.$$

Thus the components of W_n are the time elapsed since the last record occurred and the excess of the respective record value over the trend value. Also, the chains (K_n, X_n) and W_n are related by $X_m = W_{K_m}^{(2)} + cK_m$.

This leads us to the next theorem.

Theorem 7.4.2.2: (i) The sequence $\{W_n: n \geq 1\}$ is a homogeneous Markov chain with transition probabilities

$$P\left(W_{n+1}^{(1)} = k, W_{n+1}^{(2)} > y \mid W_{n1}^{(1)} = j, W_n^{(2)} = x\right)$$

$$= \begin{cases} 1(x > y) F(x - c(j + 1)) & \text{if } k = j + 1, \, j \geq 0, \\ \overline{F}(\max\{x - c(j + 1), y\}) & \text{if } k = 0, \, j \geq 0, \end{cases}$$

where $1(\cdot)$ is the indicator function.

(ii) The chain $\{W_n; n \geq 0\}$ is ergodic iff $E(Z_1^+) < \infty$, then. In that case, the stationary distribution of the chain $\pi_j(x) = \lim_{n \to \infty} P(W_n^{(1)} = j, W_n^{(2)} \leq z)$ satisfies the following equations.

$$\pi_0(\infty) = p(c),$$

$$\pi_0(z) = \sum_{m \geq 0} \int_{-\infty}^{z + c(m+1)} (F(z) - F(y - c(m + 1))) \left[\prod_{j=1}^{m} F(y - cj)\right] \pi_0(dy),$$

and

$$\frac{d\pi_m}{d\pi_0}(y) = \prod_{j=1}^{m} F(y - cj).$$

(iii) If, for some $s > 0$, $E(Z_1^+)^{s+1} < \infty$, then the total variation distance is

$$\sup_A \left| P(W_n \in A) - \sum_j \int_{A_j} \pi_j(dw) \right| = O(n^{-s}), \quad A_j = \{w \in \mathbb{R}: (j, w) \in A\}.$$

If $E\, e^{sZ_1^+} < \infty$, then the right-hand side of the above equation can be replaced by $O(e^{-sn})$.

(iv) If $E\, Z_1^+ = \infty$, then $W_n \to (\infty, \infty)$ in probability as $n \to \infty$. Moreover, if

$$\int_0^\infty \exp\left\{-(1/c)\int_0^x \overline{F}(y)dy\right\}dx<\infty,$$

then $W_n^{(2)}\to\infty$ a.s. as $n\to\infty$.

7.5 The "General Record Model"

Here, we discuss a completely different approach to describing record-breaking data. As mentioned at the beginning of this chapter, Carlin and Gelfand (1993) recommend a completely general record model where the only information available to the researcher is the joint distribution of the records and the corresponding record times. Maximum likelihood estimation for this model is difficult, however, since this approach leads to likelihood functions that contain high-dimensional integrals, rendering closed-form solution impossible. As a result, Monte Carlo methods are used to achieve approximations and then the approach is applied to model Olympic high-jump data. We now describe this method.

To enumerate the likelihood, the same notation as Carlin and Gelfand (1993) is adopted here. Let the original set of measurements be given by $\mathbf{Y}=(Y_1, Y_2, \ldots, Y_n)$. Let $1=s_1<s_2<\cdots<s_r\leq n$ denote the record times; that is, we have r records within a sequence of n events. The upper record value sequence then is denoted by $Y_{s_1}<Y_{s_2}<\ldots<Y_{s_r}$ and the data set is given by $\{Y_1,s_2,Y_{s_2},s_3,\ldots,s_r,Y_{s_r}\}$. One further assumes that the Ys have a joint density function given by $f(y;\,\theta)$, where θ is a vector of unknown parameters. The likelihood function is denoted as $L(\theta,Y_1,s_2,Y_{s_2},s_3,\ldots s_r,Y_{s_r})$. This likelihood is then written as

$$f(y_1;\theta)P(s_2|y_1,\theta)f(y_{s_2}|y_1,s_2,\theta)\cdots P(s_r|y_1,s_2,\ldots,y_{s_{r-1}};\theta)\,f(y_{s_r}|y_1,s_2,\ldots,y_{s_{r-1}},s_r;\theta)$$

$$\text{x } P(\text{no records after } t_{s_r}|y_1,s_2,\ldots,y_{s_{r-1}},s_r;\theta). \tag{7.5.1}$$

Define $\mathbf{U}=(Y_1, Y_2, \ldots, Y_{s_r})$ and let $\mathbf{V}=\mathbf{Y}/\mathbf{U}$. Then we can rewrite $f(y;\theta)$ as $f(y;\theta)=f(u,v,\theta)=f(u;\theta)f(v\mid u;\theta)$. Now define the events A_i by

$$A_i=\left\{Y_{s_i+1}\leq y_{s_i},\ldots,Y_{s_{i+1}-1}\leq y_{s_i}\right\},\ i=1,2,\ldots,r,$$

and with

$$\{V \in B\} \equiv \left\{\bigcap_{j=1}^{r} A_j\right\},$$

the likelihood (7.5.1) can be rewritten as

$$\int_B f(u,v;\ \theta)\ dv\ =\ f(u;\ \theta)\ P(V \in B \mid u;\ \theta). \qquad (7.5.2)$$

Assuming that the Ys form a Markov sequence, the above can be rewritten as

$$f(y_1;\ \theta)\ \left\{\prod_{j=2}^{r} f(y_{s_j} \mid y_{s_{j-1}};\ \theta)\right\}\ \left\{\prod_{j=2}^{r-1} P(A_j \mid y_{s_j}, y_{s_{j-1}};\ \theta)\right\}\ P(A_r \mid y_{s_r};\ \theta). \qquad (7.5.3)$$

Maximizing the above likelihood will require $(n - r)$-dimensional integration over a constrained region and is not available explicitly. In fact, typically, $(n - r)$ will be large and one will not even be able to approximate the above integral, unless the original sample consists of independent observations, in which case we will have $(n - r)$ one-dimensional integrals. As a result, Carlin and Gelfand (1993) used an iterative Monte Carlo approach to find the maximum likelihood estimates. The approach involves the creation of a Monte Carlo approximant for (7.5.3) (using the methods of Geyer and Thompson, 1992) and maximizing the resulting approximate likelihood function. An additional iterative step ensures that the likelihood itself is maximized. This approach allows the analysis of a number of parametric models, for instance, conditionally independent hierarchical models, moving window sum processes, and the Markov—Gaussian models.

Carlin and Gelfand (1993) also used this technique to fit a model to record-breaking Olympic high-jump data since 1896, as presented in the *World Almanac and Book of Facts* (1989). The data are enumerated in Table 7.5.1. Note that the data are not only a prototype for many sports history data sets, but also contain two important types of missing data. First, there were no record-breaking high jumps in the years 1904, 1920, 1928, 1932, 1948, 1972, and 1984. Second, no record occurred in the war years, that is, 1916, 1940, and 1944. The likelihood function, therefore, had to be able to make the distinction between failures and cancellations.

A Gaussian AR(1) model was fit to the data for the following reasons. First of all, a model for dependent data seems plausible here since athletes often compete in more than one year's Olympics and the fact that particular events are often dominated by the same countries.

Second, there were only 14 records set in the 21 Olympics held, suggesting perhaps a mean that is increasing over time (this makes sense intuitively, also).

A Gaussian linear AR(1) model for the record Y_i is

$$Y_i - \mu_i = \rho\,(Y_{i-1} - \mu_{i-1}) + \varepsilon_i,$$

where $\varepsilon_i \sim N(0,\ \sigma^2)$, and where for simplicity, Carlin and Gelfand (1993) let the means be defined as $\mu_i = \alpha + \beta(\text{year} - 1892)/4$.

Now, writing $\underline{w} = (y_9,\ y_{10})'$, $\underline{v} = (y_3,\ y_{20},\ y_{23},\ y_7,\ y_{14},\ \underline{w}')'$ and letting $\theta = (\alpha,\ \beta,\ \sigma,\ \rho)$, the likelihood function is given by

$$L(\theta;\ u_0) = f(y_1;\ \theta)\left\{\prod_{j=2}^{r} f(y_{s_j}\,|\,y_{s_{j-1}};\ \theta)\right\}\left\{\prod_{j\in F_1}\int_{-\infty}^{y_{j-1}} f(y_j\,|\,y_{j-1},\,y_{j+1};\ \theta)dy_j\right\}$$

$$\times\left\{\int_{-\infty}^{y_5} f(y_7\,|\,y_5,y_8;\ \theta)\,dy_7\right\}\left\{\int_{-\infty}^{y_{11}} f(y_{14}\,|\,y_{11},y_{15};\ \theta)\,dy_{14}\right\}$$

$$\times\left\{\int_{-\infty}^{y_8}\int_{-\infty}^{y_8} f(\underline{w}\,|\,y_8,y_{11};\ \theta)\,d\underline{w}\right\}, \tag{7.5.4}$$

where $F_1 = \{3,\ 20,\ 23\}$. All of the distributions above are univariate normal distributions except for that of $\underline{w}$, which is expressed as a bivariate normal.

In order to maximize the above likelihood and use it for prediction, the seven values at which records were not set need to be estimated. All failures except for those in 1928 and 1932 represented gaps of length one and were derived from conditional distributions using standard multi variate normal theory. When a failure abutted a cancellation, as in 1920, some modifications had to be made to the mean and variance to reflect that the adjacent record was more than one position away, but the calculations were still routine. For gaps of length two, Gibbs sampling was used to estimate the missing values. With the missing values in hand, the approximant to the likelihood function was generated. Parametric bootstrap was then used for the prediction of future records from 1992 until 2016. The fitted model and the predictive probabilities are shown in Table 7.5.2, which gives the maximum likelihood estimates, and Table 7.5.3, which gives the predictive probabilities.

Table 7.5.1. Olympic High-Jump Records, 1896–1988

j	s_j	Year	Record (in.)	Athlete (Country)
1	1	1896	71.25	Ellery Clark (US)
2	2	1900	74.80	Irwing Baxter (US)
3	4	1908	75.00	Harry Porter (US)
4	5	1912	76.00	Alma Richards (US)
5	8	1924	78.00	Harold Osborne (US)
6	11	1936	80.00	Cornelius Johnson (US)
7	15	1952	80.32	Walter Davis (US)
8	16	1956	83.50	Charles Dumas (US)
9	17	1960	85.00	Robert Shavalakadze (USSR)
10	18	1964	85.75	Valery Brumel (USSR)
11	19	1968	88.25	Dick Fosbury (US)
12	21	1976	88.50	Jacek Wszoia (Poland)
13	22	1980	92.75	Gerd Wessig (E. Germany)
14	24	1988	93.50	Guennadi Avdeenko (USSR)

Table 7.5.2. Maximum Likelihood Estimation for the AR(1) Model

Quantity	Estimate	Standard Error
α	70.04	1.38
β	0.89	0.08
σ	1.79	0.37
ρ	0.33	0.37

Table 7.5.3. Parametric Bootstrap Prediction for the Olympic High-Jump Data

Quantity (Year)	Bootstrap Probability that the Next Record Occurs During the Year
1992	0.39
1996	0.25
2000	0.19
2004	0,11
2008	0.05
2012	0.01
2016	0.00

As we can see, there are several reasonable models for record-breaking data when it is assumed that the underlying population is undergoing change. These models have been developed over a long period of time, but there is still much room for improvement in order to better describe the record phenomena and to be able to make more accurate prediction of future records. The challenge is there!

References

Ahsanullah, M. (1980). Linear Prediction of Record Values for the Two Parameter Exponential Distribution. *Annals of the Institute of Statistical Mathematics*, **32**, 363–368.

Ahsanullah, M. (1993). Record Values of Univariate Distributions. *Pakistan Journal of Statistics*, **9**, 49–72.

Ahsanullah, M. (1995). *Record Statistics*. Nova Science Publishers, Commack, New York.

Andersen, P.K. and Gill, R.D. (1982). Cox's Regression Model for Counting Processes: A Large Sample Study. *Annals of Statistics*, **10**, 1100–1120.

Arnold, B.C., Balakrishnan, N., and Nagaraja, H.N. (1998). *Records*. John Wiley & Sons, New York.

Balakrishnan, N. and Chan, P.S. (1994). Record values from Rayleigh and Weibull distributions and associated inference. *NIST Special Publication 866, Proceedings of the Conference on Extreme Value Theory and Applications, Vol. 3*, (Eds., J. Galambos, J. Lechner, and E. Simiu), 41–51.

Balakrishnan, N. and Chan, P.S. (1998). On the Normal Record Values and Associated Inference. *Statistics and Probability Letters*, **39**, 73–80.

Balakrishnan, N. and Cohen, A.C. (1991). *Order Statistics and Inference: Estimation and Methods*. Academic Press, San Diego.

Balakrishnan, N., Ahsanullah, M., and Chan, P.S. (1995). On the Logistics Record Values and Associated Inference. *Journal of Applied Statistical Science*, **2**, 233–248.

Ballerini, R. and Resnick, S.I. (1985). Records from an Improving Population. *Journal of Applied Probability*, **22**, 487–502.

Ballerini, R. and Resnick, S.I. (1987). Records in the Presence of a Linear Trend. *Advances in Applied Probability*, **19**, 801–828.

Basak, P. and Bagchi, P. (1990). Applications of Laplace Approximation to Record Values. *Communications in Statistics, Part A—Theory and Methods*, **19**, 1875–1888.

Berger, M. and Gulati, S. (2000). Record-Breaking Data: A Parametric Comparison of the Inverse-Sampling and the Random-Sampling Schemes. *Journal of Statistical Computation and Simulation*, **69**, 225–238.

Berred, A.M. (1992). On Record Values and the Exponent of a Distribution with a Regularly Varying Upper Tail. *Journal of Applied Probability*, **29**, 575–586.

Berred, A.M. (1998). Prediction of Record Values. *Communications in Statistics—Theory and Methods*, **27**, 2221–2240.

Borokov, K. (1999). On Records and Related Processes for Sequences with Trends. *Journal of Applied Probability*, **36**, 668–681.

Carlin, B.P. and Gelfand, A.E. (1993). Parametric Likelihood Inference for Record-Breaking Problems. *Biometrika*, **80**, 507–515.

Chan, P.S. (1998). Interval Estimation of Location and Scale Parameters Based on Record Values. *Statistics and Probability Letters*, **37**, 49–58.

Chandler, K.M. (1952). The Distribution and Frequency of Record Values. *Journal of the Royal. Statistical Society, Ser. B* **14**, 220–228.

Csörgő, M. (1983). Quantile Processes with Statistical Applications. *CBMS-NSF Regional Conference in Applied Mathematics*, SIAM, Philadelphia.

Dunsmore, I.R. (1983). The Future Occurrence of Records. *Annals of the Institute of Statistical Mathematics*, **35**, 267–270.

Dwass, M. (1960). Some *k*-Sample Rank Order Tests, in *Contributions to Probability and Statistics*, 198–202, Stanford University Press, Stanford, California.

Dwass, M. (1964). Extremal Processes. *Annals of Mathematical Statistics*, **35**, 1718–1725.

Foster, F.C. and Stuart, A. (1954). Distribution-Free Tests in Time Series Based on the Breaking of Records. *Journal of the Royal Statistical Society, Ser. B*, **16**, 1–22.

Feuerverger, A. and Hall, P. (1996). On Distribution-Free Inference for Record-Value Data with Trend. *Annals of Statistics*, **24**, 2655–2678

Galambos, J. (1978). *The Asymptotic Theory of Extreme Order Statistics*. John Wiley & Sons, New York.

Galambos, J. (1987). *The Asymptotic Theory of Extreme Order Statistics, Second Edition*. Kreiger, Malabar, Florida.

Geyer, C.J. and Thompson, E.A. (1992). Constrained Monte Carlo Maximum Likelihood for Dependent Data (with discussion). *Journal of the Royal Statistical Society, Ser. B*, **55**, 657–699.

Glick, N. (1978). Breaking Records and Breaking Boards. *American Mathematical Monthly*, **85**, 2–26.

Gnedenko, B. (1943). Sur la Distribution Limited u Terme Maximum d'une Serie Aleatoire. *Annals of Mathematics*, **44**, 423–453.

Goldberger, A.S. (1962). Best Linear Unbiased Predictors in the Generalized Regression Model. *Journal of the American Statistical Association*, **57**, 369–375.

Gulati, S. (1991). *Smooth Nonparametric Estimation from Record-Breaking Data*. Ph.D. Dissertation, University of South Carolina, Columbia.

Gulati S. and Padgett, W.J. (1992). Kernel Density Estimation from Record-Breaking Data, in P. Goel and J. Klein (Eds.), *Survival Analysis: State of the Art*. Kluwer Academic, The Netherlands, 197–210.

Gulati, S. and Padgett, W.J. (1994a). Smooth Nonparametric Estimation of Distribution and Density Functions from Record-Breaking Data. *Communications in Statistics, Theory and Methods*, **23**, 1259–1274.

Gulati, S. and Padgett, W.J. (1994b). Nonparametric Quantile Estimation from Record-Breaking Data. *Australian Journal of Statistics*, **36**, 211–223.

Gulati, S. and Padgett, W.J. (1994c). Smooth Nonparametric Estimation of the Hazard and the Hazard Rate Functions from Record-Breaking Data. *Journal of Statistical Planning and Inference*, **42**, 331–341.

Gulati, S. and Padgett, W.J. (1994d). Estimation of Nonlinear Statistical Functions from Record-Breaking Data: A Review. *Nonlinear Times and Digest*, **1**, 97–112.

Gulati, S. and Padgett, W. (1995). Nonparametric Function Estimation from Inversely Sampled Record-Breaking Data. *The Canadian Journal of Statistics*, **23**, 359–368.

Hoel, P.G., Port, S.C., and Stone, C.J. (1972). *Introduction to Stochastic Processes.* Houghton Mifflin, Boston.

Hoinkes, L.A., and Padgett, W.J. (1994). Maximum Likelihood Estimation from Record-Breaking Data for the Weibull Distribution. *Quality and Reliability Engineering International*, **10**, 5–13.

Kaplan, E.L. and Meier, P. (1958). Nonparametric Estimation from Complete Observations. *Journal of the American Statistical Association*, **53**, 457–481.

Kingman, J.F.C. (1973). Subadditive Ergodic Theory. *Annals of Probability*, **1**, 883–909.

Mann, N.R. (1969). Optimum Estimators for Linear Functions of Location and Scale Parameters. *Annals of Mathematical Statistics*, **40**, 2149–2155.

Mengoni, L. (1973). *World and National Leaders in Track and Field Athletics, 1860–1972*. Ascoli Piceno, Italy.

Nadaraya, E.A. (1965). On Nonparametric Estimation of Density Functions and Regression Curves. *Theory of Probability and its Applications*, **10**, 186–190.

Nagaraja, H. (1988). Record Values and Related Statistics—A Review. *Communications in Statistics, Theory Methods*, **17**, 2223–2238.

Nagaraja, H.N. (1984). Asymptotic Linear Prediction of Extreme Order Statistics. *Annals of the Institute of Statistical Mathematics*, **36**, 289–299.

Neuts, M.F. (1967). Waiting Times Between Record-Breaking Observations. *Journal of Applied Probability*, **4**, 206–208.

Nevzorov, V.B. (1985). Records and Interrecord Times for Sequences of Nonidentically Distributed Random Variables. *Zapiski Nauchn. Semin. LOMI*, **142**, 109–118 (in Russian). Translation in *Journal of Soviet Mathematics*, **36** (1987), 510–516.

Nevzorov, V.B. (1987). Records. *Theory of Probability and Applications*, **32**, 221–228.

Nevzorov, V.B. (1990). Records for Nonidentically Distributed Random Variables, in B. Grigelionis, Yu.V. Prohorov, V.V. Sazano, and V. Statulevicius (Eds.), *Proceedings of the 5ᵗʰ Vilnius Conference*,Vol. **2**. VSP, Mokslas, 227–233.

Nevzorov, V.B. (1995). Asymptotic Distribution of Records in Nonstationary Schemes. *Journal of Statistical Planning and Inference*, **45**, 261–273.

Parzen, E. (1962). On Estimation of a Probability Density Function and Mode. *Annals of Mathematical Statistics*, **33**, 1065–1076.

Pfeifer, D. (1982). Characterizations of Exponential Distributions by Independent Non-Stationary Record Increments. *Journal of Applied Probability*, **19**, 127–135 (Correction. **19**, p. 906)

Proschan, F. (1963). Theoretical Explanation of Observed Decreasing Failure Rate. *Technometrics*, **5**, 375–383.

El-Qasem, A.A. (1996). Estimation via Record Values. *Journal of Information and Optimization Sciences*, **17**, 541–548.

Renyi, A. (1962). Théorie des Elements Saillants d'une Suite d'Observations, with Summary in English. *Colloquium on Combinatorial Methods in Probability Theory*, 104–117, Mathematisk Institut, Aarhus Universitet, Denmark.

Resnick, S.I. (1973a). Limit Laws for Record Values. *Stochastic Processes and Their Applications*, **1**, 67–82.

Resnick, S.I. (1973b). Record Values and Maxima. *Annals of Probability*, **1**, 650–662.

Resnick, S.I. (1973c). Extremal Processes and Record-Value Times. *Journal of Applied Probability*, **10**, 864–868.

Resnick, S.I. (1987). *Extreme Values, Regular Variation, and Point Processes*. Springer-Verlag, New York.

Robinson, M.E. and Tawn, J.A. (1995). Statistics for Exceptional Records. *Applied Statistics*, **44**, 499–511.

Rosenblatt, M. (1956). Remarks on Some Nonparametric Estimates of a Density Function. *Annals of Mathematical Statistics*, **27**, 832–835.

Samaniego, F.J. and Kaiser, L.D. (1978). Estimating Value in a Uniform Action. *Naval Research Logistics Quarterly*, **25**, 621–632.

Samaniego, F.J. and Whitaker, L.R. (1986). On Estimating Population Characteristics from Record-Breaking Observations. I. Parametric Results. *Naval Research Logistics Quarterly*, **33**, 531–543.

Samaniego, F.J. and Whitaker, L.R. (1988). On Estimating Population Characteristics from Record-Breaking Observations. II: Nonparametric Results. *Naval Research Logistics Quarterly*, **35**, 221–236.

Shorack, G.R. and Wellner, J.A. (1986). *Empirical Processes with Applications to Statistics*. John Wiley & Sons, New York.

Shorrock, R.W. (1972a). A Limit Theorem for Inter-Record Times. *Journal of Applied Probability*, **9**, 219–223.

Shorrock, R.W. (1972b). On Record Values and Record Times. *Journal of Applied Probability*, **9**, 316–326.

Shorrock, R.W. (1973). Record Values and Inter-Record Times. *Journal of Applied Probability*, **10**, 543–555.

Smith, R.L. (1988). Forecasting Records by Maximum Likelihood. *Journal of the American Statistical Association*, **83**, 331–338.

Smith, R.L. (1997). Statistics for Exceptional Records (Letter to the Editors). *Applied Statistics*, **46**, 123–128.

Smith, R.L. and Miller, J.E. (1986). A Non-Gaussian State Space Model and Applications to the Prediction of Records. *Journal of the Royal Statistical Society, Ser. B*, **48**, 79–88.

Tata, M.N. (1969). On Outstanding Values in a Sequence of Random Variables. *Zeitschrift fuer Wahrscheinlichkeitstheorie und Verwandte Gebiete*, **12**, 9–20.

Tierney, L. and Kadane, J.B. (1986). Accurate Approximations for Posterior Moments and Marginal Densities. *Journal of the American Statistical Association*, **81**, 82–86.

Tiwari, R.C. and Zalkikar, J.N. (1991). Bayesian Inferences of Survival Curve from Record-Breaking Observations: Estimation and Asymptotic Results. *Naval Research Logistics Quarterly*, **38**, 599–609.

Tryfos, P. and Blackmore, R. (1985). Forecasting Records. *Journal of the American Statistical Association*, **80**, 46–50.

Vervaat, W. (1972). Functional Central Limit Theorem for Processes with Positive Drift and Their Inverses. *Zeitschrift fuer Wahrscheinlichkeitstheorie und Verwandte Gebiete*, **23**, 245–253.

Weissman, I. (1978). Estimation of Parameters and Large Quantiles Based on the k Largest Observations. *Journal of the American Statistical Association*, **73**, 812–815.

World Almanac and Book of Facts (1989). Newspaper Enterprise Association, New York.

Yang, M.C.K. (1975). On the Distribution of the Inter-Record Times in an Increasing Population. *Journal of Applied Probability*, **12**, 148–154.

Yang, S.S. (1985). A Smooth Estimator of a Quantile Function. *Journal of the American Statistical Association*, **80**, 1004–1011.

Index

Ahsanullah, M., 7, 11, 25, 26, 28, 29, 30, 105
almost surely, 9, 59, 89, 91
alternate hypothesis, 36
Andersen, P., 43, 105
Arnold, B.C., 3, 4, 7, 9, 12, 25, 28, 30, 31, 36, 68, 81, 82, 84, 87, 105
asymptotically normal, 3, 8, 9, 34, 35, 45, 60, 89
athletic records, 28, 92, 97

Bagchi, P., 67, 69, 70, 72, 105
Balakrishnan, N., 11, 12, 25, 26, 27, 28, 31, 105
Ballerini, R., 3, 82, 87, 89, 92, 105
bandwidth, 45, 46, 47, 48, 52, 60
Basak, P., 67, 69, 70, 72, 105
Bayes estimator, 64, 73, 74, 75, 78, 79
Bayesian inference, 31, 65, 67, 73
Bayesian nonparametric inference, 33
Bayesian predictive distributions, 11
Berger, M., 19, 20, 105
Berred, A.M., 31, 32, 105
best linear invariant predictor, 11, 25, 28
best linear unbiased predictor, 11, 25
Blackmore, R., 11, 28, 108
Borokov, K., 82, 97, 98, 105

Carlin, B.P., 82, 100, 101, 102, 105
Chan, P.S., 12, 25, 31, 105
Chandler, K.M., 2, 3, 7, 8, 9, 105
classical record model, 2, 5, 7, 28, 81, 83, 84, 87
consistency, 14, 43, 52, 55, 57, 58, 59, 60, 97
covariance function, 44, 55, 56, 59, 62, 76
Csörgő, M., 56, 105

density estimator, 45, 46, 47, 49, 52, 63
Dirichlet process, 67, 73, 76
distribution-free tests, 3, 4, 11
Dunsmore, I.R., 11, 25, 30, 67, 68, 106
Dwass, M., 2, 7, 9, 106

empirical Bayes, 67, 73, 78, 79, 80
empirical hazard function, 48
empirical quantile function, 47
exponential decay model, 96
exponential distribution, 3, 11, 12, 14, 15, 21, 23, 25, 31, 38, 61, 62, 67, 68, 72, 85, 93
extreme value distribution, 11, 19, 28, 29, 93, 95

failure stress, 5
failure times, 15, 16, 61
Feuerverger, A., 82, 96, 97, 106
Foster, F.C., 4, 11, 33, 34, 35, 36, 82, 106
Fourier inversion, 35
future records, 11, 25, 33, 82, 102, 104
F^α model, 81, 83, 84

Galambos, J., 7, 9, 30, 105, 106
gamma distribution, 14, 24, 32, 69, 72, 93
Gaussian process, 44, 51, 52, 54, 55, 56, 59, 60, 76
Gelfand, A.E., 82, 100, 101, 102, 105
general record model, 82, 100
geometric distribution, 13
geometric model, 81, 82, 83, 84, 85, 86, 87
Gill, R., 43, 105
Glick, N., 5, 6, 7, 8, 106
Goldberger, A.S., 25, 29, 106
Guinness, 1
Gulati, S., 3, 19, 20, 38, 46, 47, 59, 60, 61, 62, 105, 106
Gumbel distribution, 31, 95

Hall, P., 82, 96, 97, 106
hazard function, 40, 43, 46, 48, 59, 70, 71
hazard rate function, 48, 49, 59, 61, 63
high-jump data, 100, 101, 103
Hoel, P., 55, 107
Hoinkes, L., 17, 18, 19, 64, 107

interrecord time sequence, 7, 9, 84
inverse sampling scheme, 6, 12, 19, 20, 64

Kaiser, L.D., 11, 108
Kaplan—Meier estimator, 37, 55
kernel function, 45, 46, 47, 48, 53, 59, 60
Kingman, J.F.C., 91, 107

Laplace approximation, 67, 69, 70, 71, 72
likelihood function, 13, 14, 17, 22, 23, 36, 95, 100, 101, 102
linear drift model, 96
linear predictors, 25
linear trend model, 81, 82, 96
location-scale family, 25, 31
logistic distribution, 12, 25, 26, 27, 28
lognormal distribution, 61
loss function, 67, 73, 78
lower records, 28, 33, 34, 89, 95

marathon records, 93, 94, 95, 96
Markov chain, 9, 67, 82, 88, 97, 98, 99
maximum likelihood estimator, 3, 12, 15, 36, 39, 74
mean-squared error, 18, 19, 20, 37, 46, 60, 61
Miller, J.E., 92, 93, 94, 96, 108
minimum variance unbiased estimator, 14, 15
mixture, 15
mode, 7, 9, 71
moment generating function, 13

Nadaraya-type estimator, 48, 49, 58

Nagaraja, H.N., 7, 11, 12, 25, 30, 105, 107
Neuts, M.F., 7, 9, 107
Nevzorov, V.B., 7, 82, 83, 87, 107
nonparametric function estimation, 4
nonparametric inference, 3, 4, 33, 67
nonparametric maximum likelihood estimation, 12, 33, 36
normal distribution, 8, 12, 28, 31, 32, 61, 95, 96, 102
null hypothesis, 34
number of records, 6, 7, 8, 9, 14, 19, 32, 89, 91

Olympic games, 84, 87
Olympic records, 83, 87, 89
optimal predictors, 29
order statistic, 1, 22, 30

Padgett, W.J., 3, 17, 18, 19, 46, 47, 59, 60, 61, 62, 64, 106
Pareto distribution, 19, 20, 21, 22, 23, 25, 30
Parzen, E., 45, 46, 107
percentile, 5, 7, 32, 36, 95
Pfeifer, D., 81, 83, 84, 85, 87, 88, 107
Port, S., 55, 107
posterior distribution, 67, 69, 70
prediction, 1, 3, 4, 11, 24, 25, 28, 30, 31, 32, 33, 35, 67, 68, 69, 82, 92, 93, 94, 95, 102, 103, 104
prediction intervals, 30, 31, 67
predictive distribution, 11, 67, 68, 69, 70, 71, 72, 93
prior distribution, 69, 70, 71
Proschan, F., 15, 61, 74, 79, 107

Qasem, A.A. El-, 21, 24, 25, 107
quadratic-drift model, 96
quantile process, 55

random sampling scheme, 6, 12, 15, 19, 20
randomly right-censored data, 37
record time sequence, 7, 9

record value sequence, 7, 8, 9, 82, 84, 89, 95, 100
relative efficiency, 22, 23, 24, 38, 64
Renyi, A., 2, 7, 9, 107
Resnick, S., 3, 7, 8, 9, 82, 87, 89, 92, 105, 107
Robinson, M.E., 97, 108
Rosenblatt, M., 45, 108

Samaniego, F.J., 3, 4, 11, 12, 13, 14, 15, 16, 17, 33, 36, 37, 39, 43, 44, 48, 50, 53, 73, 74, 76, 78, 108
shock model, 83
Shorack, G., 55, 108
Shorrock, R.W., 7, 9, 108
Smith, R.L., 3, 82, 92, 93, 94, 95, 96, 97, 108
smooth function estimation, 33, 45
smoothing parameter, 61
state space model, 92, 93
stationary distribution, 99
Stone, C., 55, 107
strong law of large numbers, 7, 43, 77, 78
strong uniform consistency, 43, 48, 49, 50, 55, 57, 58, 60
strong uniform consistent estimator, 41, 43
Stuart, A., 4, 11, 33, 34, 35, 36, 82, 106
sufficient statistic, 14, 68
survival function, 12, 32, 39, 37, 67, 73, 74, 76, 78, 79, 80

Tawn, J.A., 97, 108
Tiwari, R.C., 4, 67, 73, 74, 75, 76, 78, 79, 108
tolerance limit, 5, 6
tolerance regions, 11, 31, 32
trend, 2, 3, 4, 33, 34, 36, 81, 82, 83, 84, 89, 94, 95, 96, 97, 98
Tryfos, P., 11, 28, 108

uniform distribution, 11, 21, 22, 23, 28
upper records, 11, 21, 33, 36, 67, 69, 88, 89

variance-covariance matrix, 25, 27

Weibull distribution, 12, 17, 18, 19, 20, 21, 25, 30, 33, 38, 60, 64, 72
Weissman, L., 11, 25, 108
Wellner, J., 55, 108
Whitaker, L., 3, 4, 12, 13, 14, 15, 16, 17, 33, 36, 37, 39, 43, 44, 48, 50, 53, 73, 74, 76, 78, 108
wooden beam, 1, 5

Yang, M.C.K., 3, 82, 83, 84, 86, 87, 89, 108
Yang, S.S., 58, 108

Zalkikar, J., 4, 67, 73, 74, 75, 76, 78, 79, 108

Lecture Notes in Statistics

For information about Volumes 1 to 120, please contact Springer-Verlag

121: Constantine Gatsonis, James S. Hodges, Robert E. Kass, Robert McCulloch, Peter Rossi, and Nozer D. Singpurwalla (Editors), Case Studies in Bayesian Statistics, Volume III. xvi, 487 pp., 1997.

122: Timothy G. Gregoire, David R. Brillinger, Peter J. Diggle, Estelle Russek-Cohen, William G. Warren, and Russell D. Wolfinger (Editors), Modeling Longitudinal and Spatially Correlated Data. x, 402 pp., 1997.

123: D. Y. Lin and T. R. Fleming (Editors), Proceedings of the First Seattle Symposium in Biostatistics: Survival Analysis. xiii, 308 pp., 1997.

124: Christine H. Müller, Robust Planning and Analysis of Experiments. x, 234 pp., 1997.

125: Valerii V. Fedorov and Peter Hackl, Model-Oriented Design of Experiments. viii, 117 pp., 1997.

126: Geert Verbeke and Geert Molenberghs, Linear Mixed Models in Practice: A SAS-Oriented Approach. xiii, 306 pp., 1997.

127: Harald Niederreiter, Peter Hellekalek, Gerhard Larcher, and Peter Zinterhof (Editors), Monte Carlo and Quasi-Monte Carlo Methods 1996. xii, 448 pp., 1997.

128: L. Accardi and C.C. Heyde (Editors), Probability Towards 2000. x, 356 pp., 1998.

129: Wolfgang Härdle, Gerard Kerkyacharian, Dominique Picard, and Alexander Tsybakov, Wavelets, Approximation, and Statistical Applications. xvi, 265 pp., 1998.

130: Bo-Cheng Wei, Exponential Family Nonlinear Models. ix, 240 pp., 1998.

131: Joel L. Horowitz, Semiparametric Methods in Econometrics. ix, 204 pp., 1998.

132: Douglas Nychka, Walter W. Piegorsch, and Lawrence H. Cox (Editors), Case Studies in Environmental Statistics. viii, 200 pp., 1998.

133: Dipak Dey, Peter Müller, and Debajyoti Sinha (Editors), Practical Nonparametric and Semiparametric Bayesian Statistics. xv, 408 pp., 1998.

134: Yu. A. Kutoyants, Statistical Inference For Spatial Poisson Processes. vii, 284 pp., 1998.

135: Christian P. Robert, Discretization and MCMC Convergence Assessment. x, 192 pp., 1998.

136: Gregory C. Reinsel, Raja P. Velu, Multivariate Reduced-Rank Regression. xiii, 272 pp., 1998.

137: V. Seshadri, The Inverse Gaussian Distribution: Statistical Theory and Applications. xii, 360 pp., 1998.

138: Peter Hellekalek and Gerhard Larcher (Editors), Random and Quasi-Random Point Sets. xi, 352 pp., 1998.

139: Roger B. Nelsen, An Introduction to Copulas. xi, 232 pp., 1999.

140: Constantine Gatsonis, Robert E. Kass, Bradley Carlin, Alicia Carriquiry, Andrew Gelman, Isabella Verdinelli, and Mike West (Editors), Case Studies in Bayesian Statistics, Volume IV. xvi, 456 pp., 1999.

141: Peter Müller and Brani Vidakovic (Editors), Bayesian Inference in Wavelet Based Models. xiii, 394 pp., 1999.

142: György Terdik, Bilinear Stochastic Models and Related Problems of Nonlinear Time Series Analysis: A Frequency Domain Approach. xi, 258 pp., 1999.

143: Russell Barton, Graphical Methods for the Design of Experiments. x, 208 pp., 1999.

144: L. Mark Berliner, Douglas Nychka, and Timothy Hoar (Editors), Case Studies in Statistics and the Atmospheric Sciences. x, 208 pp., 2000.

145: James H. Matis and Thomas R. Kiffe, Stochastic Population Models. viii, 220 pp., 2000.

146: Wim Schoutens, Stochastic Processes and Orthogonal Polynomials. xiv, 163 pp., 2000.

147: Jürgen Franke, Wolfgang Härdle, and Gerhard Stahl, Measuring Risk in Complex Stochastic Systems. xvi, 272 pp., 2000.

148: S.E. Ahmed and Nancy Reid, Empirical Bayes and Likelihood Inference. x, 200 pp., 2000.

149: D. Bosq, Linear Processes in Function Spaces: Theory and Applications. xv, 296 pp., 2000.

150: Tadeusz Caliński and Sanpei Kageyama, Block Designs: A Randomization Approach, Volume I: Analysis. ix, 313 pp., 2000.

151: Håkan Andersson and Tom Britton, Stochastic Epidemic Models and Their Statistical Analysis. ix, 152 pp., 2000.

152: David Ríos Insua and Fabrizio Ruggeri, Robust Bayesian Analysis. xiii, 435 pp., 2000.

153: Parimal Mukhopadhyay, Topics in Survey Sampling. x, 303 pp., 2000.

154: Regina Kaiser and Agustín Maravall, Measuring Business Cycles in Economic Time Series. vi, 190 pp., 2000.

155: Leon Willenborg and Ton de Waal, Elements of Statistical Disclosure Control. xvii, 289 pp., 2000.

156: Gordon Willmot and X. Sheldon Lin, Lundberg Approximations for Compound Distributions with Insurance Applications. xi, 272 pp., 2000.

157: Anne Boomsma, Marijtje A.J. van Duijn, and Tom A.B. Snijders (Editors), Essays on Item Response Theory. xv, 448 pp., 2000.

158: Dominique Ladiray and Benoît Quenneville, Seasonal Adjustment with the X-11 Method. xxii, 220 pp., 2001.

159: Marc Moore (Editor), Spatial Statistics: Methodological Aspects and Some Applications. xvi, 282 pp., 2001.

160: Tomasz Rychlik, Projecting Statistical Functionals. viii, 184 pp., 2001.

161: Maarten Jansen, Noise Reduction by Wavelet Thresholding. xxii, 224 pp., 2001.

162: Constantine Gatsonis, Bradley Carlin, Alicia Carriquiry, Andrew Gelman, Robert E. Kass Isabella Verdinelli, and Mike West (Editors), Case Studies in Bayesian Statistics, Volume V. xiv, 448 pp., 2001.

163: Erkki P. Liski, Nripes K. Mandal, Kirti R. Shah, and Bikas K. Sinha, Topics in Optimal Design. xii, 164 pp., 2002.

164: Peter Goos, The Optimal Design of Blocked and Split-Plot Experiments. xiv, 244 pp., 2002.

165: Karl Mosler, Multivariate Dispersion, Central Regions and Depth: The Lift Zonoid Approach. xii, 280 pp., 2002.

166: Hira L. Koul, Weighted Empirical Processes in Dynamic Nonlinear Models, Second Edition. xiii, 425 pp., 2002.

167: Constantine Gatsonis, Alicia Carriquiry, Andrew Gelman, David Higdon, Robert E. Kass, Donna Pauler, and Isabella Verdinelli (Editors), Case Studies in Bayesian Statistics, Volume VI. xiv, 376 pp., 2002.

168: Susanne Rässler, Statistical Matching: A Frequentist Theory, Practical Applications, and Alternative Bayesian Approaches. xviii, 238 pp., 2002.

169: Yu. I. Ingster and Irina A. Suslina, Nonparametric Goodness-of-Fit Testing Under Gaussian Models. xiv, 453 pp., 2003.

170: Tadeusz Caliński and Sanpei Kageyama, Block Designs: A Randomization Approach, Volume II: Design. xii, 351 pp., 2003.

171: David D. Denison, Mark H. Hansen, Christopher C. Holmes, Bani Mallick, Bin Yu (Editors) Nonlinear Estimation and Classification. viii, 474 pp., 2003.

172: Sneh Gulati and William J. Padgett, Parametric and Nonparametric Inference from Record-Breaking Data. viii, 111 pp., 2003.